Embodied Minds – Technical Environments

Edited by Thomas Hoff and Cato A. Bjørkli

Embodied Minds – Technical Environments
Conceptual Tools for Analysis, Design and Training

tapir academic press

Tapir Academic Press, Trondheim 2008

ISBN 978-82-519-2341-5

This publication may not be reproduced, stored in a retrieval system or transmitted in any form or by any means; electronic, electrostatic, magnetic tape, mechanical, photo-copying, recording or otherwise, without permission.

Layout: Type-it AS, Trondheim
Printed by Tapir Uttrykk
Binding: Grafisk Produksjonsservice AS

This book has been published with founding from NTNU´s Interdisciplinarity project "Mobilitetens Tekniske Rom".

In norwegian:

Denne boken er støttet av NTNU's tverrfaglige prosjekt «Mobilitetens Tekniske Rom».

Tapir Academic Press
7005 TRONDHEIM
Tlf.:   73 59 32 10
Faks: 73 59 84 94
E-post: forlag@tapir.no
www.tapirforlag.no

Contents

List of Contributors 7

Introduction 11
Cato A. Bjørkli and Thomas Hoff

Part I: Systems Analysis 23
1 Working with Human Errors: Concepts and Applications 25
Salvatore Massaiu

2 A Formative Analysis of the Train Driving Task 75
Synve Røine

3 Railway Driving Operations and Cognitive Ergonomics Issues in the Norwegian Railway: A Systems Analysis 113
Sarah M. Brotnov

Part 2: Interface Design 145
4 Ecological Interaction Properties 147
Thomas Hoff and Kjell Ivar Øvergård

5 On Representations of In-Vehicle Information Systems – Effects of Graphical (GUI) versus Speech-Based User Interfaces 161
Cato A. Bjørkli, Thomas Hoff, Gunnar D. Jenssen, Trond A. Øritsland and Pål Ulleberg

6 Theory and Practice of Ecological Interfaces: A Case Study of a Haptic In-Vehicle Audio System Design 181
Thomas Hoff, Hans Vanhauwaert Bjelland and Cato A. Bjørkli

Part III: Simulator Skills Training 219
7 The Role of Fidelity, Transfer and Cognitive Involvement in Learning – A review of simulator training 221
Paul Andreas Lundeby

8 User-Centred Development of Simulator-based Training – An Exploratory Case Study 237
Kjell-Morten Bratsberg Thorsen

9 A Video-Based Phenomenological Method for Evaluation of Driving Experience in Staged or Simulated Environments 259
Kjell Ivar Øvergård

List of Contributors

Thomas Hoff holds a doctorate in psychology and works as Associate Professor of Organizational Psychology and Innovation Studies at the Department of Psychology at the University of Oslo. His research focuses on the interplay between humans and their immediate surroundings in an ecological psychology perspective, and the implications of this theoretical position for aspects of human–technology interaction, organizational psychology and innovation studies. Hoff has published papers in journals such as *The Design Journal, Ergonomics, Theoretical Issues in Ergonomics Science and Cognition*, and *Technology and Work*. In 2008, he co-edited the anthology *Spaces of Mobility* published by Equinox Publishing. He is currently affiliated with the Centre for Advanced Study at the Norwegian Academy of Science and Letters. In the project 'Understanding Innovation' he works on the topic of organizational antecedents of innovation capability in firms.

Cato Alexander Bjørkli is a clinical psychologist with a doctorate in psychology, and has several international publications on the theme of human–technology interaction. His main fields of interest extend across philosophical, methodological, theoretical, and practical implications of the relationship between humans and technology. His key areas of competence are modelling of human performance in complex systems, methodology for the study of human performance in complex systems, and design and innovation. In 2005–2007 Bjørkli worked at Sintef Technology and Society, in Trondheim, as researcher, and participated in several national and international research projects related to development of intelligent driver support systems, HMI challenges for vehicle information systems, and simulator training and evaluation. Today, Cato Bjørkli holds a position as Associate Professor of Work and Organisational Psychology at the University of Oslo, Norway.

Kjell Ivar Øvergård holds a master's degree in psychology from the Department of Psychology at the Norwegian University of Science and Technology (NTNU). He currently holds a position as a doctoral research fellow at the same department. He has several publications in international journals in the fields of cognitive systems engineering, human factors, ergonomics, and interaction design. His research interests include embodied cognition, control modelling as the integration between perception, cognition and action, human technology interaction, and the quantitative modelling of human cognition.

Salvatore Massaiu holds a graduate degree in philosophy from the University of Pisa and an MSc in Marketing Research from the Norwegian School of Management BI, in Oslo. He works as research scientist at the Industrial Psychology Department of the Institute of Energy Technology – OECD Halden Reactor Project. He has been conducting research in the areas of nuclear power plant simulator experiments, human reliability, human factors test and evaluation, and transport safety.

Trond Are Øritsland has an MA in Industrial Design, with a doctoral degree in engineering. He works as an Associate Professor at the Department of Product Design, at the Norwegian University of Science and Technology (NTNU). He teaches industrial design, man–machine interaction, and graphic user interface design. His research interests lie within the field of user-centred design (UCD) and interaction design. He is interested in the application of embodied mind theory in industrial design thinking and is currently exploring the problems of distance collaboration and shared visual displays across disciplines. From 2008 he is the supervisor of three PhDs within the Center for Integrated Operations in the Petroleum Industry, at NTNU in Trondheim. Øritsland is currently working on his second textbook focussing on man–machine interaction. He has previously published *A Theory of Discursive Interaction Design: Mapping the Development of Quality and Style in Man Machine Interaction* (1999).

Synve Røine holds a master's degree in psychology. She currently works as usability specialist at Netlife Research, in Oslo. She has taught human-technology interaction courses at the University of Oslo and has carried out research within the train domain. Her key areas of competence are human performance in complex systems and information technology usability. Røine's main interest is the study and analysis of human performance in complex systems and she incorporate these topics in her daily work with usability in information technology.

Pål Ulleberg is an Associate Professor of Psychometrics at the Department of Psychology, University of Oslo. He has worked within the field of traffic psychology since 1999, especially focusing upon the relationship between personality, attitudes, driving behaviour, and accident involvement. During recent years he has primarily carried out experiments and field studies estimating the effect of different road safety measures on driver workload, behaviour and accidents. He has also carried out studies examining the predictive validity of tests measuring visual and functional impairment on driving behaviour.

Paul Andreas Lundeby is Director of Product and Development at Simoveo AS, in Oslo. He holds an MPh in Psychology from the University of Oslo, and his primary area of interest is the relationship between learning in virtual environments and the effect of transfer. He is currently leading the research and development of a negotiation simulator at Simoveo AS. Paul Andreas also holds courses on negotiation, with emphasis on the psychological aspects of negotiation in relation to game-based learning.

Sarah M. Brotnov holds an MPh in Psychology from the University of Oslo. She currently works as a principal advisor at the Norwegian Railway Inspectorate, in Oslo. She works with all human factors related issues connected to railway infrastructure and technology development within the Norwegian railway. The Inspectorate directs its efforts towards ensuring that rail traffic is operated in a safe and appropriate manner in the best interests of passengers, rail company employees and the general public. The Inspectorate is responsible for ensuring that rail operators meet the conditions and requirements set out in rail legislation that governs the traffic. The authority is also responsible for drawing up regulations, awarding licences for rail activity and approving rolling stock and infrastructure. Brotnov has carried out several research projects within the human factors discipline, including a large project on maritime pilots in cooperation with the Institute of Transport Economics.

Hans Vanhauwaert Bjelland holds a doctoral degree in engineering. He is currently employed as Assistant Professor at the Department of Product Design, Norwegian University of Science and Technology (NTNU). His project 'Touching Technology' addresses the theoretical and technological downplay of touch in human interaction with technology. The study explores how designing for human physical interaction with technology involves an understanding of touch, how it takes part in interaction, and the technological constraints of using touch-related technologies. Additionally, the project studies how this knowledge can be incorporated in the design process. Bjelland's main fields of interests focus on the intersection between theoretical and practical issues related to human interaction with technology and the design of new products.

Kjell-Morten Bratsberg Thorsen works as a Usability Specialist at NetLife Research in Oslo. He holds an MPhil in Psychology from the University of Oslo. He also has a background in web development and programming. His main field of interest is user-centred development and in particular, user experience of the Internet.

Gunnar Deinboll Jenssen is currently a Senior Research Scientist at SINTEF Transport Safety and Informatics, in Trondheim. In 2003–2007 Gunnar Jenssen was engaged in PhD work focusing on safety effects and behavioural adaptation to advanced driver support systems (ADAS) He obtained his MSc in Psychology in 1986, at The University of Trondheim (later NTNU). Until 1989 Gunnar Jenssen was a Research Fellow at the Institute of Psychology (NTNU), with research on perception, and learning disabilities. In 2005 he was appointed as an expert on fitness to drive by the European Commission DG Energy and Transport Driving Licence Committee. He has conducted several simulator studies related to development and evaluation of intelligent transportation systems (ITS) and HMI solutions for in-vehicle information systems based on visual tactile and speech interfaces. Gunnar Jenssen has extensive experience in international collaborative work from participation as a project manager and task manager in the EU Framework projects PROMETHEUS, DRIVE, PROMPT, IMMORTAL, STARDUST, UPTUN, and CLARESCO, and also as a project manager of two (three-year) US-funded research programmes on effectiveness of road signs and rail signs.

Introduction

Cato A. Bjørkli and Thomas Hoff

The synthesis of human operators and technology into functional systems has a rather short history, but an elaborate past. From our earliest origins, humans have applied tools and primitive technologies, but the last hundred years have shown a remarkable development of the relationship between humans and artefacts. Flach (2000) outlines the short, but condensed development during the past century where the human capacity for work has undergone vast changes. The first efforts in modern time to coordinate humans and machines emphasized the rationalization of human effort through detailed work procedures in pair with mechanical machines (Taylor 1911).

However, the perspectives on human-technology interaction were challenged as the nature of work changed from primarily physical labour to more cognitive mediated work. World War II spurred a 'technological sprint' involving the rapid development of weapons and defences, in which airplanes exemplified a qualitatively different setting for human workers. The pilots were embedded in a technical context (the aircraft and cockpit) that placed demands on the subjective understanding of the system in quite a different way than workers at the assembly lines of the Ford factories. The key to efficiency and skilled operation shifted from physical strength to cognitive skills. Accidents and incidents seldom originated in lack of physical strength, but in lack of understanding of the system – exemplified in accidents where fully functional airplanes crashed into the ground. The research paradigm of American researchers Fitts and Jones established human performance and technology as a defining theme of human factors (Fitts and Jones 1947).

Since the era of Fitts and Jones, the last fifty years of technological development have accentuated integration and entwinement of humans and technology into unitary syntheses where vast physical forces and informational streams are coordinated into complex socio-technical systems (Dörner 1996; Casey 1998; Perrow 1999). The classic examples of modern complex systems are control rooms, air traffic control, electric grid management, off-shore installations, stock exchanges, and nuclear power compounds.

Similarly, the theories and perspectives on these complex human-technology systems have changed in many ways. It has become clear that we have created systems that hold qualities not foreseen in formal engineering and design. There is a widespread consensus that the characterization of complex systems includes descriptions of system behaviour that emphasize the non-linear and dynamic aspects of system operation (Woods 1988; Rasmussen 1997; Vicente 1999; Hollnagel and Woods 2005; Woods and Hollnagel 2006). The patterns of work in complex systems reveal features as abrupt changes in tempo and rhythm of operation, intermingled and cascading effects, progressive escalation from routine to non-routine situations, unexpected coupling between subcomponents, automation surprises, uncertain data sets, and changing temporal constraints (Norros 2004; Woods and Hollnagel 2006).

A central feature of modern societies is the deep integration of Joint Cognitive Systems (JCS) in the constitution and functioning of social structures such as economy, transport, military operations, and politics (Kallinikos 2005). There is thus an equally widespread dependence on safe, efficient and productive operation of JCS in society. Hence, the modern human-technology unity based on the constellation of operators and artefacts yields functional entities that demand extensive coordinative and regulative measures to be taken. However, the challenge is that the establishment of coordination and control in complex systems evades much of the theories provided by classic psychology and/or engineering (Vicente 1999; Hollnagel 2004). The explanatory deficiency often becomes lucid in accident investigations of failed operation of complex systems in which the inclusion of characteristic features of the interaction between the system and context in question is difficult to come to terms with (Dekker 2005; Le Coze, Salvi and Gaston 2006). Here, the 'modern accidents' indicate that the relationship between humans and technology in complex systems has changed drastically since the early studies of workers at the assembly line. We now are faced with entities that override our previous most basic assumptions of what divides humans from machines. Complex socio-technical systems may thus be understood as the articulation of the profound question of how humans relate to our surrounding world.

It seems that the clear-cut distinction between the living organisms and material objects is at least debatable. A wide range of studies within anthropology, neurobiology, and zoology of tool use in both human and non-human organisms reveals that material objects have a twilight status – they are, on one hand, a material part of the world in terms of having physical extension. At the same time, they are deeply integrated in the human experience of the world (Ingold 2000; Johnson-Frey 2004; Laeng, Brennen, Johannesen, Holmen and Elvestad 2004; Maravita and Iriki 2004; Lewis 2006). We live with and through objects. These studies indicate that the relationship between an organism and tools is somewhat open-ended, in the sense that the use of tools is not just moving physical matter about (e.g. wielding a stick) – it appears that tool use entails some form of integration of the tool into the body schemata of the organism and further that it structures the way the organism relates to the world (e.g. a blind person orienting his/herself).

However, the emergence of complex socio-technical systems represents an elaborate form of tool use, whereby technology and tools become not just single material objects used by a person, but entire technological contexts through which the organisms (read: workers) experience the world. This suggests a deep integration of humans in technology (and vice versa). Here, the capacities of the organism are extended into the immediate objects and interfaces, but at the same time the world also changes character.

The modern complex systems, such as the control rooms and airplanes, and offshore installations, represent the construction of technological contexts that removes the operator from an immediate and direct presence in the world into a state of elaborate mediation. The technological context becomes an 'experiential veil' between the human and the world (Ihde 1983; Mitcham 1994). However, the nature of this 'veil' is somewhat dual. Norros (2004) argues that work characterized by technological mediation becomes both more physically remote, but at the same time invites the operator deeper into the object in terms of richer information and abstraction. This suggest that technology enters between the operator and the given environment where work takes place and mediates the relevant features of it by way of formal representations and higher order relationships. For example, traditional radars represent the surroundings in simplified 2D graphics, and infrared cameras reduce the range of colours perceived but enhance visual perception in poor light conditions. Here, the environment takes the form of representations in terms of numbers and simplified shapes, thus reducing and removing the operator from the context. Yet, the environment becomes richer in terms of information (for more examples of higher order representations, see Burns and Hajdukiewicz (2004). Here, technology opens up the context in new ways as the operator experiences the world *through* technology.

This 'Heideggerian' line of reasoning emphasizes the active engagement in the world through tools, where the non-perceivable context becomes perceivable (Fløistad 1993; Ingold 2000; Dreyfus and Wrathall 2005). Technological mediation may here be understood as a form of restructuring of the experience of the environment for the operator – giving rise to a new form of habitat for humans. Complex socio-technical systems then emerge as novel technical habitats constructed for some specific purpose (e.g. power management, oil production, economic trade). Researchers within the field of cognitive engineering (CE) may thus be regarded as occupied with the ecology of human-technology systems as they focus on the challenges associated with the introduction of new technology into complex socio-technical systems (Flach, Hancock, Caird and Vicente 1994; Vicente 1994; Woods and Dekker 2001; Hoff, 2004).

Vicente (1994) discuss the consequences of regarding complex socio-technical systems as ecological systems, and argues that one of the main implications is that the unit of analysis in human factors changes from the behaviour and characteristics of single operators to the reciprocally coupled human-technology system. Central to the coupling of humans and technology into a functional system is the description of the environment with relevance to system goals – what are the features of the surrounding that bear relevance to the goal at hand? Here, the concepts of *affordances* and *constraints*[1] are central to establishing the reciprocal coupling (Rasmussen, Pejtersen and Goodstein 1994; Ingold 2000; Albrechtsen, Andersen, Bødker and Pejtersen 2001; Vicente 2003). These concepts serve to emphasize the connection between a controlled process and the human-technology habitat in question. The functional entity of a goal-oriented human-technology system unfolds in correspondence with some external physical reality (e.g. flow electricity in electrical grids, underground oil pressure during drilling operations, wind and weather during flying, etc.). These external (or environmental) constraints exist independently of and apart from any belief or preconception of the operators.

Complex socio-technical systems may thus be regarded as 'correspondence-driven systems' as they adapt to an outside process (Vicente 1999). In contrast, coherence-driven defines the human-technology dyads that do not have to take an external reality into account – for example, gaming consoles, driving simulators, board games, and computerized word processors (Microsoft Word). In the closed, coherence-driven systems, the fit between the user and the artefact can be fine tuned in accordance with the cognitive and perceptual characteristics of the user. In correspondence-driven systems, however, there is a source of constant disturbances and variation in system performance. In this case, there are two sets of constraints to respect: the environmental constraints and the cognitive/perceptual constraints of the operators.

The reason for spending so much effort in distinguishing between the unit of analysis and its relationship to the context is rooted in the need to attribute the complexity and origin of patterns correctly. Simon (1981) described a now widely cited example of the problem of tracing the generating mechanism for the movement patterns of an ant walking about on a beach. The example illustrates that a detailed geometrical description of the path of the ant suggest a significant complex pattern and widespread irregularity. To look for the generating mechanism of this complexity within the ant (the ant brain, preferably) is arguably a faulty attribution of the origin of complexity in which the contextual features (surface and terrain of the beach) are left out.

Beyond this rhetorical example, other researchers have shown this principle empirically. Thelen and Smith (1994) summarized several studies of motor development in children and non-human organisms. The intriguing process of both human and non-human organisms learning to walk left researchers searching for detailed 'motor programmes' located in the host organism, either in form of cognitive schemata (Zelazo 1984), genetic coding (Forssberg 1985) or neural maturation (Konner 1991). Conversely, Thelen and Smith (1994) described several studies where the change of context conditions was found to change the motor skill level and granularity of walking movement in organisms. For example, in human motor learning, babies held torso-deep in water showed stepping movements long before being able

---

[1] The concepts *affordances* and *constraints* are used here with reference to their use within human factors research, as exemplified by Flach (1994), Albrechtsen et al. (2001) and Norros (2004). See the cited references for a full discussion of the definition and content of the terms.

to walk and thus forestalling their present motor developmental stage. Further, the babies failed to display stepping movements when weights were added to their lower legs and joints in the same water condition. When held on a treadmill, the babies showed coordinated walking movements seemingly far ahead of their motor development. In non-human motor learning, studies have shown movements in hatching chickens that are characterized by synchronous thrusting with both legs. When newly hatched, this specific type of movement normally disappears and is not expressed again. However, the hatching movement may be elicited by subjecting the chicken to conditions similar to the hatching situation. Subjugated to a treadmill, 'motor-premature' chickens, cats and frogs all showed proper coordinated locomotive movement.

The interesting implication of the motor studies described by Thelen and colleagues is that movement does not seem to originate from some complex inner structure, but results from the dialogue between the biological and biomechanical constitution of the moving organism and the context specifics. The layout of limbs and joints will display advanced coordination 'prematurely' given sufficient conditions. The underlying argument is that *behavioural output should not be isolated from the behavioural situation.*

The link between hatching chickens and operator performance in complex socio-technical systems may admittedly seem somewhat far-fetched. However, the human factor researcher Kim Vicente comments on the similarity between motor control development and complex socio-technical systems as follows:

> [T]hese parallels lead to viewing the control of the human motor system and the control of complex sociotechnical systems as both involving the coordination of many degrees of freedom by a resource-limited agent (or agents.) (Vicente 1999, p. 113)

The emphasis on coordination of degrees of freedom requires some form of description of the context structures and agents involved. We have so far argued for the change of unit of analysis from the single operator to functional human-technology systems, and offered the concepts of affordance and constraints as examples of operative context denotations. The implicit line of reasoning here aims at specifying the 'limbs and joints' of complex systems and the relevant context when describing complex and coordinated behaviour in elaborate human-technology systems. In this respect, several authors have raised the issue of the coordination of the degrees of freedom rising from the operation of a complex socio-technical system in a goal-relevant context as the key feature of human factors research (Hollnagel 1992; Norros and Savioja 2002; Jagacinski and Flach 2003).

Vicente (1994) imports the argument of Thelen and Smith (1994) into human factors by stating that the pattern-generating mechanisms must be understood only in reference to the contextual layout. He quotes Neisser:

> If we do not have a good account of the information that perceivers are actually using, our hypothetical models of their information processing are almost sure to be wrong. If we do have such an account, however, such models may turn out to be almost unnecessary. (Neisser 1987)

Still, the explanatory value of perceptual information or field descriptions diminishes when the organism is faced with unfamiliar and unanticipated situations and/or underspecified perceptual information. Here, cognitive processes becomes more prominent. To the extent information and access to the context is irregular and uncertain, performance is not confined to perception and action in relation to a structured context. Uncertainty calls for reflection and understanding as basis for performance. This seems to be the case for complex socio-technical systems – operators are constantly subjected to uncertainty, dynamism and complexity (Vicente 1999; Norros 2004; Woods and Hollnagel 2006).

This brings us to the theme of how operators manage to stay in control of a dynamic process that is constantly challenged by disturbances and uncertainties at the front stage of human factors research. Further, the description of systems behaviour, operator adaptation, and multiple frames of references for work practice become important areas of interest.

This perspective marks a changing paradigm for understanding human-technology systems. It completes the transition from the 'single operator' focus accompanied by the disintegrated view on performance to the Joint Cognitive Systems approach (Hollnagel and Woods 2005). This perspective on human-technology interaction leaves us with the notion that the relationship between humans and technology in modern societies is a multifaceted and fundamentally complex phenomenon. The nature of our use of and dependency on technology is somewhat cloaked in our efforts to understand, and it resists many of our efforts to control and wilfully design systems that behave in full correspondence with our hopes and expectations. This book addresses this intricate relationship between humans and technology from many angles, and hopefully brings ideas, knowledge, and practical value to our ongoing efforts to create functional and liveable conditions for our society.

Part I of this book is devoted to the issue of unit of analysis in complex socio-technical systems. In order to understand the mechanisms involved in such systems, the task, operator and context need to be scrutinized analytically. The methods used for gaining an analytical understanding of the system need to be geared towards change. It is not sufficient merely to *describe* the different components of the system, because such a description would necessarily also encompass bad habits, workarounds, technical equipment that does not fit the task, and so forth, which would give a faulty picture of the system with respect to change. These types of methods are often referred to as formative or predictive (Vicente, 1999), as opposed to merely descriptive approaches.

Salvatore Massaiu approaches the issue of unit of analysis from the perspective of 'human error'. In line with the transition from the 'single operator' focus accompanied by the disintegrated view of performance in a joint cognitive systems approach, Massaiu argues that human error is a term that historically leans heavily towards cognition. This model is both at risk of wrongly making a scapegoat of the operator, and of creating an inappropriate basis for changing the system. Human error, Massaiu argues, is not a meaningful term unless one refers to the intentions of the operator, the intentions of the system, and the broader organizational system in which the activity take place. His paper points to areas of misunderstanding and difficulty related to current definitions of the term, particularly the reliance on the concept of intention, as well as showing the existence of a continuum between the commonly used notions of mistake and violation. He goes on to highlight the major theories or models of human performance, and examines how they treat the concept of human error and how they classify and work with the concept. In effect, Massaiu sets the stage for a systems perspective on human performance in general, and with respect to human error in particular.

Sarah Brotnov presents a practical example of a systems analysis. She aims to provide a general system safety account of the Norwegian railway. In line with the suggestions from Massaiu, Brotnov focuses her analysis on the interaction between technology, artefacts, humans, and the organization as such. The study was based on ethnographic studies, interviews, as well as an analysis of 542 reports of accidents and anomalies found in official records. As opposed to formalized methods, Brotnov makes a pragmatic and systemic attempt to pinpoint obvious shortcomings of the Norwegian railway system. Her method is mainly descriptive, as opposed to formative, as she takes the current state of affairs into consideration. However, she demonstrates that it is possible to conduct an analysis based on complex theoretical assumptions about cognitive systems, within a complicated and complex domain, and still come up with very practical suggestions in a relatively short time frame.

Synve Røine sets out to conduct an analysis of the task of train driving. Borrowing arguments for going beyond normative and descriptive accounts of work practice from Vicente (1999), Røine presents a formative approach to the analysis. The conceptual distinctions central in the formative approach are adapted from Vicente (1999), but a conceptual framework is added to describe the different levels with respect to relevant factors. This facilitates an increase in the granularity of the analysis that outlines the demands and resources at the core of railway transport. The paper concurs with the general formative aim to describe work in complex socio-technical systems as a balance between demands and the resources available. Røine proceeds to discuss the use of the formative approach and to what extent it actually solves the insufficiencies of traditional work analysis methods that initially motivated the formulation of a formative approach. The conclusion reached is that a purely formative work analysis is at least problematic, due to the fact that it somehow builds on a descriptive analysis of how the system works today. Røine goes on to argue that how the actual analysis of descriptive data is formulated might be the key to a formative view on work practice in complex socio-technical systems.

Part II of this book relates to the implementation of changes in technology based on an analysis of operators, tasks and contexts. In particular, this section is devoted to user interfaces, i.e. the part of a technical system that the operator interacts with. In the history of Human-Computer Interaction (HCI), fairly little is said about interface design per se, especially regarding the ecological qualities of interfaces (representations in the interface that are coherent with evolution-based sensitivity toward certain types of information). In principle, it probably impossible to state principles and guidelines that encompass all situations, no matter what the context of use is (Øvergård, Fostervold, Bjelland and Hoff 2007). However, there are properties of the human perceptual apparatus that make us more sensitive towards some types of information than toward others. In this section, a theoretical paper, an experiment-based paper and a design case paper approach this topic from several angles.

Thomas Hoff and Kjell Øvergård present a framework termed Ecological Interaction Properties (EIP), where they aim to derive some principles for interaction design based on a theoretical standpoint in turn derived from ecological psychology and embodied mind philosophy. This structural typology points towards elements of the user interface that give rise to an altered phenomenological perception of the interaction with that interface. This is based on Hoff, Øritsland and Bjørkli (2002), who show that it is possible to make radical technological changes to an interface without really changing the experience of interacting with the artefact. The EIP properties aim to be both a conceptual tool for designing user interfaces, but also act as an evaluative tool, as carried out by, among others, Røed, Gould, Bjørkli and Hoff (2005), who evaluated the transition from the old, hardwired bridge of the Hauk class high-speed craft of the Royal Norwegian Navy, to the digitalized 'class cockpit' of the new Skjold class.

Cato Bjørkli, Thomas Hoff, Gunnar Jenssen, Trond Are Øritsland and Pål Ulleberg present a car simulator study of how different interfaces in vehicle information systems affect driver performance. The consequences of introducing such systems in cars are somewhat unknown beyond a vague assumption that they might negatively affect driver performance. The paper aims to formulate a theoretical basis for understanding and predicting the consequences of what happens when a driver becomes involved in a secondary task of this level of complexity during driving. The paper presents a simulator study that compares the use of a commercial information system accessed by two different interfaces. The outcome is measured in terms of subjective workload (NASA-TLX) and objective measures (speed and vehicle positioning). The results indicate that the speech-recognition system shows less negative impact on driver performance than the manually operated graphical interface. However, the difference is perhaps not as marked as the theoretical basis suggests. The authors discuss which aspects of our understanding of driver performance are most useful in guiding the design and evaluation of future information systems in vehicles.

Thomas Hoff, Hans Vanhauwaert Bjelland and Cato Bjørkli present a case study where the principles of Ecological Interaction Properties (EIP) are exploited in the case of in-car audio systems for personal vehicles. Here, the authors aim to demonstrate that in order to create effective, safe and fun user-interfaces, one needs to study both the context of use (compliant with the user-centred design tradition, represented by e.g. ISO 13407) as well as principles of ecological interface design, as represented by the EIP categories. In order to give the readers a feel for the development of a user interface design process, sketches, mock-ups, and prototypes are presented in the paper. In contrast to Bjørkli, Hoof, Jenssen, Øritsland and Ulleberg (2008, this volume) the aim is not to demonstrate empirically that this user interface is superior to others, but rather to show how design principles might be used in a case study.

Part III of this book is devoted to simulator-based learning. Training is by far the most effective means for compensating for poor human-technology interactions. As a complementary approach to systems design, this topic thus deserves consideration. Simulator training is a type of training that is particularly relevant to the topic of this book, namely complex technological environments. Simulated environments have the advantage of letting the operator explore the boundaries of system performance without any operational risk. Traditionally, simulator training has focussed on procedure training, or rule-based behaviour (Vicente 1999), and to some extent towards knowledge-based behaviour (anomaly detection and corrective actions). The issue of skill-based training, however, has received little consideration in the research literature. All three papers in this section of the book are devoted to this matter. The first paper addresses skill-based simulator training from a theoretical and conceptual point of view, the next paper presents a case study of the development of a skill-based simulator, while the final paper in this section presents a novel methodological approach for use in different types of simulators.

Paul Lundeby sets out to explore how simulators and games have been conceptualized in the literature. He then goes on to classify different types of simulator training according to Rasmussen's (1983) skill-, rules- and knowledge-based behaviour. In addition, Lundeby touches on the issue of fidelity. To what extent is does fidelity (the real-world like qualities of the simulator) determine the quality of the simulator training? As a guide for the design of skill-based simulators, Lundeby presents a number of inherent tensions that the design team needs to take into consideration.

Whereas Lundeby discuss the theoretical foundation for skill-based simulators, and points at some directions for simulator design, Kjell-Morten Bratsberg Thorsen treats the issue of simulator design explicitly. Which design methods should be used in order to create effective skill-based simulators? His central claim is that most User Centred Design (UCD) methods are directed towards the development of tools that are intended to support the operator in his/her work. The purpose of a training simulator, however, is not to support work but to simulate it as an environment for training and exploration. Hence, the development of a training simulator relates to two distinct contexts of use – one that will be simulated and offered in training, and one in which the actual training will occur. Bratsberg Thorsen's aim is to explore the effect of utilizing principles of UCD and Contextual Design through a case study of the development of a negotiation skill simulator.

Kjell Ivar Øvergård's paper completes the book and reflects on how the dominant perspectives within traffic safety research have emphasized quantification of driver performance. This precedence of quantified measures of traffic behaviour is exemplified in three distinct traditions in traffic safety research: 1) engineering approach, 2) traffic psychology, and 3) psychophysiology. Øvergård argues that these traditional approaches could be supplemented by qualitative data describing the driver experience in order to increase the effects of traffic safety measures. The research strategy for acquiring data is presented within the tradition of phenomenology, in which the description of subjective experience is the preferred focus of analysis. From a description of the fundamental phenomenological principles, Øvergård outlines a research strategy to triangulate qualitative and quantitative data, and argues that such a combination allows deeper insight into how drivers respond to traffic environments.

The papers in this book investigate different aspects of how humans use technology, and each paper adopts a different approach to addresses the issue of how to improve and control technology use. In our modern world, technology remains the two-edged sword that both brings the future closer and pushes us backwards to making the same old mistakes. The elusive character of our relationship to the world around us blurs the boundary between us and the tools we use. The deep integration of technology into modern societies continues to challenge our efforts to understand how best to shape our future through the worlds we create for ourselves and the coming generations.

References

Albrechtsen, H., Andersen, H. H. K., Bødker, S. and Pejtersen, A. M. (2001). *Affordances in Activity Theory and Cognitive Systems Engineering*. Roskilde: RISØ National Laboratory.
Burns, C. M. and Hajdukiewicz, J. R. (2004). *Ecological Interface Design*. Boca Raton, FL: CRC Press.
Casey, R. (1998). *Set Phasers on Stun: And Other True Tales of Design, Technology, and Human Error*. Santa Barbara, CA: Aegan.
Dekker, S. W. (2005). Ten Questions About Human Error, Lawrence Erlbaum Associates, Inc.
Dörner, D. (1996). *The Logic of Failure*. New York: Basic Books.
Dreyfus, H. L. and Wrathall, M. A. (eds.) (2005). *A Companion to Heidegger*. Victoria: Blackwell.
Fitts, P. M. and Jones, R. E. (1947). *Analysis of Factors Contributing to 460 Pilot-error Experiences in Operating Aircraft Controls*. Dayton, Oh: Aero Medical Laboratory, Wright-Patterson Air Force Base.
Flach, J. M. (1994). The ecology of human-machine systems: A Personal History. In: J. M. Flach, Hancock, P., Caird, J. and Vicente, K. J. (Eds.): Global Perspectives on the Ecology of Human-Machine Systems. Hillsdale, NJ, LEA: 1-13.
Flach, J. M. (2000). Human capacity for work: A (biased) historical perspective. In P. Hancock and Desmond, P. A. (eds.) *Stress, Workload, and Fatigue: Theory, Research, and Practice*, 429–442. Mahwah, NJ: Erlbaum.
Flach, J. M., Hancock, P., Caird, J. and Vicente, K. J. (eds.) (1994). *Global Perspectives on the Ecology of Human-Machine Systems: A Global Perspective (Resources for Ecological Psychology*. Vol. 1. New York: LEA.
Fløistad, G. (1993). *Heidegger: En innføring i hans filosofi*. Oslo: Pax Forlag.
Forssberg, H. (1985). Ontogeny of human motor locomotor control. I. Infant stepping, supported locomotion, and transition to independent locomotion. *Experimental Brain Research* 57: 480–493.
Hoff, T. (2004). "Comments on the Ecology of Representations in Computerised Systems." Theoretical Issues in Ergonomics Science 5(5): 453-472.
Hoff, T., Øritsland, T. A. and Bjørkli, C. A. (2002). Ecological interaction properties. *NordDesign* 2002: 137–151.

Hollnagel, E. (1992). Coping, Coupling, and Control: The modelling of muddling through. *Mental Models and Everyday Activities. Proceedings of the 2nd Interdiscipinary Workshop on Mental Models, Robinson College, Cambridge, 23–25 March 1992*, 61–73.

Hollnagel, E. (2004). *Barriers and Accident Prevention*. Aldershot: Ashgate.

Hollnagel, E. and Woods, D. D. (2005). *Joint Cognitive Systems: Foundations of Cognitive Systems Engineering*. Boca Raton, FL: CRC Press.

Ihde, D. (1983). *Existential Technics*. New York: State University Press.

Ingold, T. (2000). *The Perception of the Environment*. London: Routledge.

Jagacinski, R. J. and Flach, J. M. (2003). *Control Theory for Humans: Quantitative Approaches to Modelling Performance*. Mahwah, NJ: LEA.

Johnson-Frey, S. H. (2004). The neural bases of complex tool use in humans. *Trends in Cognitive Science* 8(2): 71–78.

Kallinikos, J. (2005). The order of technology: Complexity and control in a connected world. *Information and Organisation* 15(3): 185–202.

Konner, M. (1991). Universals of behavioural development in realtion to brain myelination. In K. R. Gibson and Petersen, A. C. (eds.) *Brain Maturation and Cognitive Development: Comparative and Cross-Cultural Perspectives*, 181–223. New York: Aldine de Gruyter.

Laeng, B., Brennen, T., Johannesen, K., Holmen, K. and Elvestad, R. (2004). Multiple reference frames in neglect? An investigation of the object-centred frame and the dissociation between 'near' and 'far' from the body by use of a mirror. *Cortex* 38: 511–528.

Le Coze, J. C., Salvi, O. and Gaston, D. (2006). Complexity and multi (inter or trans)-disciplinary sciences: Which job for engineers in risk management? *Journal of Risk Research* 9(5): 569–582.

Lewis, J. W. (2006). Cortical networks related to human use of tools. *The Neuroscientist* 12(3): 211–231.

Maravita, A. and Iriki, A. (2004). Tools for the body (schema). *Trends in Cognitive Science* 8(2): 79–86.

Mitcham, C. (1994). *Thinking through Technology: The Path between Engineering and Philosophy* Chicago: University of Chicago Press.

Neisser, U. (1987). From direct perception to conceptual structure. In U. Neisser (ed.) *Concepts and Conceptual Development: Ecological and Intellectual Factors in Organisation*, 11–24. Cambridge, MA: Cambridge University Press.

Norros, L. (2004). *Acting under Uncertainty: The Core-task Analysis in Ecological Study of Work*. Espoo: VTT

Norros, L. and Savioja, P. (2002). Modelling of the activity system – development of an evaluation method for integrated system validation. Meetings of the enlarged Halden Programme Group Meeting. Sandefjord.

Øvergård, K. I., Fostervold, K. I., Bjelland, H. V. and Hoff, T. (2007). Knobology in use: An experimental evaluation of ergonomics recommendations. *Ergonomics* 50(5): 694–705.

Perrow, C. (1999). *Normal Accidents*. Princeton NJ: Princeton University Press.

Rasmussen, J. (1983). Skill, rules, knowledge; signals, signs, and symbols, and other distinctions in human performance models. *IEEE Transactions on Systems, Man and Cybernetics* SMC-13(3): 257–266.

Rasmussen, J. (1997). Risk managment in a dynamic society: A modelling problem. *Safety Science* 27(2): 183–213.

Rasmussen, J., Pejtersen, A. M. and Goodstein, L. P. (1994). *Cogntive Systems Engineering*. New York: John Wiley.

Simon, H. A. (1981). *The Sciences of the Artifical*. Cambridge, MA: MIT Press.

Taylor, F. W. (1911). *The Principles of Scientific Managment*. New York: Harper & Bros.

Thelen, E. and Smith, L. B. (1994). *A Dynamic Systems Approach to the Development of Cognition and Action*. Cambridge MA: MIT Press.

Vicente, K. J. (1994). A few ecological implications of an ecological approach to human factors. In J. M. Flach, Hancock, P., Caird, J. and Vicente, K. J. (eds.) *Global Perspectives on the Ecology of Human-Machine Systems: A Global Perspective*. Vol. 1. *Resources for Ecological Psychology*. LEA 1: 424.

Vicente, K. J. (1999). *Cognitive Work Analysis: Toward Safe, Productive, and Healthy Computer-Based Work.* Mahwah, NJ: LEA.

Vicente, K. J. (2003). *The Human Factor: Revolutionizing the Way People Live with Technology.* New York: Routledge.

Woods, D. D. (1988). Coping with complexity: The psychology of human behaviour in complex systems. In L. P. Goodstein, Andersen, H. and Olsen, S. E. (eds.) *Tasks, Errors, and Mental Models: A Festschrift to Celebrate the 60th Birthday of Jens Rasmussen*, 128–148. London: Taylor & Francis.

Woods, D. D. (1998). Designs are hypotheses about how artifacts shape cognition and collaboration *Ergonomics* 41(2): 168–173.

Woods, D. D. and Dekker, S. W. (2001). Anticipating the effects of technology change: A new era of dynamics for human factors. *Anticipating the Effects of Technology Change: A New Era of Dynamics for Human Factors* 1(3): 272–282.

Woods, D. D. and Hollnagel, E. (2006). *Joint Cognitive Systems: Patterns in Cognitive Systems Engineering.* Boca Raton FL: Taylor & Francis.

Zelazo, P. R. (1984). The development of walking: New findings and old assumptions. *Journal of Motor Behaviour* 15: 99–137.

Part I:

Systems Analysis

1

Working with Human Errors: Concepts and Applications

Salvatore Massaiu

Human error identification, classification and reduction have long been considered activities of critical importance to system safety. However, twenty years of specialised research on the issue have indicated that the concept human error is far more complicated that originally assumed, to the point that some authors have recently proposed rejecting it altogether. This essay analytically investigates the concept by reviewing the way it is currently defined and studied within the dominating theories of human performance in safety-critical systems. The main thesis is that human errors are not fixed events that can be studied by means of observation alone. Instead, human error is a normative concept, which implies a process of comparing empirical events with abstract standards of correct performance. This, in turn, implies that both the process and the standards are dependent upon theories describing how humans behave in given situations. One consequence is that counting and classifying human errors within and across studies and domains will hardly progress safety. At the same time, awareness of the real nature of the concept is the only way to preserve a role for the concept of human error in modern safety science.

1 Human Error

It is virtually impossible to review the issue of human error without finding articles and books that report on the percentage contribution of human errors to system failures. A review of incident surveys by Hollnagel (Hollnagel, 1993) shows that the estimated contribution of 'human errors' to incidents ranges from about 20% to around 80%. The fact that the surveys covered a quite short period of time, from 1960 to 1990, makes it unlikely that such huge differences in the estimates can be explained by the transformations in the human-machine environment in those decades (see Hollnagel, 1993).

So what explains this variability? One could point to the heterogeneity of the errors counted, which typically concentrate on errors in operation but also include other phases of human-machine systems interaction, such as design, maintenance and management. A second reason can be attributed to the different industries surveyed, which in the review cited spawned from nuclear power plants to aerospace and from weapon systems to general studies. These factors do certainly explain a good deal of this variability, but even if we were to concentrate on the same industry and on one particular class of actors (i.e. front line operators), we would still find very different estimates.

The reason is that in field of human factors there is no general consensus on the meaning of the expression 'human error'. However, there are various models and perspectives on human performance that incorporate different interpretations of the concept human error. They bring about an extraordinary diversity of notions and applications that they associate with the label 'human error', and they produce incident analysis methods and classification systems of errors that are typically only partially compatible with each other.

A further problem with the expression 'human error' is that traditionally it has been associated with the attribution of responsibility and blame. In this context, human error is typically a judgement of human performance made after an event has occurred. Old views of human errors as dominant causes of accidents have influenced the disciplines of accident investigation and error analysis up to the present day, to the point that some authors have debased the label 'human error' to an 'ex post facto judgement made at hindsight' (Woods et al., 1994), with very little or no utility in advancing knowledge about system safety, or even rejected the label altogether: 'there is no such thing as human error' (Hollnagel, 1993).

I believe there is still a use for the label 'human error', provided we clearly define its meaning and delimit its applications in ways that counter the biases implicit in intuitive and traditional uses of the expression. Hence, in the following sections I will thoroughly analyse the concept of human error, highlight the areas of misunderstanding, and provide a minimal definition capable of encompassing the majority of uses and applications. I will then review the different models or paradigms of human error analysis, and discuss some examples of classification systems that these approaches have generated. In the words of David Woods (Woods et al., 1994, p. xvii), this analysis seems necessary as,

> one cannot get onto productive tracks about error, its relationship to technology change, prediction, modelling, and countermeasures, without directly addressing the various perspectives, assumptions, and misconceptions of the different people interested in the topic of human error.

1.1 The concept of human error

The concept of human error is not an easy one, for several reasons. In the first place, even limiting the attention to the area of work psychology and human factors, there are different needs and interests to be taken into account in defining human error: for instance, human error can be defined in order to identify potential threats to system safety, as it is done in human reliability analysis, or in order to identify the causes of an accident. In the former context the definition will probably concentrate on the types of actions that an operator can perform within the system and their consequences, while in the latter the focus will likely be on the causes of the human actions that were involved in the accident. A second difficulty arises as a consequence of the different approaches to the issue: while an engineer will tend to analyse human performance in terms of success and failure in the same way as for technical elements, a sociologist will describe actions and errors in the context of the socio-technical influences and constraints under which humans operate.

The most serious difficulty, however, lies in the concept itself. Human error applies to a large variety of actions (e.g. simple tasks, cognitive operations, motor-skills), it can be attributed to a host of different causes (e.g. internal constitution, external conditions, task demands, volitions), and it can be judged with different criteria (e.g. system parameters, agents intentions, social norms). Hence, it is not an easy task to include all possible conditions and fields of applications in a simple, yet general proposition.

Typically, human error is defined within the theoretical framework provided by a discipline, for a precise scope and to specific fields of application. Available definitions are then working definitions which are more or less adequate for a scope rather than correct or incorrect in abstract terms. I will, nonetheless, advance a general definition of human error, although not for the sake of a 'correct' definition but because the process will allow identifying and discussing the essential conditions of any definition, and clarifying the meaning of the concept. By providing a definition I will be able to discuss some recurrent ones and therefore appreciate their relative strengths and limitations. This discussion will furthermore make it easier to appreciate the differences and similarities between the various approaches to human error that will be analysed later.

1.2 Errors as normative statements

In order to talk of human error, some event or action associated with undesired outcomes or consequences needs to be considered. This is the case both for mundane applications of the concept of human error, such as a child that fails to report when doing additions, or for work contexts, as when a power plant operator opens the wrong valve. The important point is, however, that this plain consideration contains the two essential elements for a definition of human error: an event and a standard of correct performance. The standard of correctness defines whether the event (or action) is an error or a correct performance, and by implication whether the consequences (real or hypothetical) associated with the event are desirable or not. It is important to stress that an action is never an error or an unsafe act by itself but it is so only in comparison with a standard of correctness and a context of execution: exactly the same action can be exemplary performance in one situation and a gross mistake in another.

An error statement is thus a judgment, where a normative propriety (e.g. wrong, too much, too fast, etc.) is assigned to a set of descriptive statements of actions and conditions of executions, in virtue of there existing a relevant standard or norm in which those actions and conditions are associated in a different way than the one observed (see Table 1.1). Table 1.1 includes other typical elements of error analysis: the assumed causes of the actions, and judgements about responsibility for the event.

Table 1.1 Types of statements involved in error analysis: two examples.

Type of statement	Example 1: Industrial process	Example 2: Road transportation
Descriptive statements		
Action	Operator A opens valve x at t	Driver A passes junction X, direction south-north at 11:32:12 pm
Conditions	Valve X is open at t_1 and $t < t_1$	Traffic light at junction X is red, direction south-north between 11:31:30 and 11:32:30
Consequences	Release of polluting substances into atmosphere	Increased risk at junction
Standard of correctness	Time t to t_1 => valve X is closed	When traffic light is red, driver stops
Error judgment	Operator A *wrongly* opens valve X at time t	Driver A *wrongly* passes junction X against red light
Causal statements		
Internal causes	Slip of action: operator A intended to open nearby valve Y but failed the execution	Perceptual confusion Circadian rhythms: sub-optimal performance at night-time
External causes	Switches of valves X and Y close in position No feedback Unavailable procedures Low lighting in room	Low visibility due to showers Input complexity: left-turn light is green
Responsibility judgment	Design of work place and working conditions are 'error forcing'	Driver A fined

The reference to norms or standards of correct performance becomes of practical relevance when dealing with actions that are not straightforwardly definable as failures: violations, performance deviations, under-specified instructions, non-procedural practices, etc. That is to say, there are practical circumstances where standards of correct performance, procedures and norms are not clearly specified and hence a preliminary discussion around them is necessary for the individuation of something as a manifestation of error. In many cases, however, there are straightforward performance criteria and it is natural to agree whether an execution has been too short, too late, on the wrong object, omitted, etc. The evaluators will have no problem in referring to the time, space and energy proprieties of an action and to characterise it in normative terms, such as for instance, wrong direction, too fast, repetition, or on wrong object.

1.3 Manifestations and causes

Hollnagel (1998) has repeatedly stressed the importance of clearly distinguishing between causes and manifestations in error classification. He claims that few of the existing error classifications make this distinction clear but mix up observable action characteristics with inferred causes. It is surely undisputable the superiority of those classifications which make the difference between error causes and manifestation explicit, and which explain how causes and manifestations are related. However, many classification systems are practical tools developed in well-defined domains where the user would not see themselves making a great deal of inference, for instance when indicating 'information communication incomplete' or 'diagnostic error'.

Further, we should not believe that error manifestations or phenotypes are mere descriptions of actions and events. An error manifestation is properly a normative statement in which the time, space and energy dimensions of an action are evaluated against an agreed standard. Clearly, when the standard is obvious the difference has no practical implications. However, as we will see later, this is not always the case. The traditional behavioural categorisation of errors in omissions and commission is a clear example of a normative process where the standards of correct performance are assumed to be clearly specified. Thus, without clearly agreed performance criteria all commissions would also be omissions of something. Furthermore, such omissions could also be described by a varied phenomenology: an action could be missing, delayed, anticipated, or replaced by another.

Whatever the performance criteria used, the evaluation process depends on the assumption that the event and the consequences are not associated by chance, but there is, instead, a causal connection (1) between the event (action or inaction) and its consequences and/or (2) between the action and the surrounding conditions that preceded it. The latter point shows that a causal explanation of some real or potential unwanted consequences ought not to stop the error manifestations, but may refer to events internal to the subject as well as external characteristics of the situation. This is the level of the causes of the manifestations and it is dependent upon the theory of behaviour underlying the explanation. Therefore, in addition to a standard of correct performance, the process of error attribution depends on the theory, or model, of human behaviour adopted. It is generally assumed that there is more than a single cause for any behaviour and that an explanation, or prediction, of a manifestation of error will include a set of causes that are deemed sufficient to have caused it, or to predict it.

1.4 Slips and mistakes

Definitions of human error are often provided from the point of view of the agent. These definitions are typically not limited to erroneous human actions or behaviours but consider mental processes as well as intentions. Mental processes (such as observation, memory, planning, etc.) are considered relevant because although they can fail without producing unwanted consequences for a particular action, they are likely to explain them on most occasions. In general, mental processes are seen as the mechanisms that underlie human actions and errors. This is a requirement of explanation, of understanding why humans do make errors, but also of description, as a limited set of causes can explain infinite erroneous actions.

Intentions, however, seem to be even more important in defining human error. A simple description of a series of events, even in the case of mental events, is insufficient to qualify them as erroneous. For instance, without reference to purposes and intentions the fact that a person did not achieve a particular goal that he/she was supposed to reach could be ascribed equally well to the person's choice of a plan of actions which was inadequate to reach his/her goal (i.e. the definition of mistake), or to his/her failure to execute an adequate plan (i.e. the definition of a slip), or to his/her purposeful selection of a goal contrary to rules and regulations (i.e. the standard definition of violation). As the example shows, there are multiple goals implied, both in the form of intentions of the agent and as intentions or standards of correctness of a group, an organisation or a system.

The mismatch between different goals and between goals and results provides the basis of a phenomenology of errors and unsafe acts. When the goals are those of a conscious actor the term 'intention' is used: that a driver had the goal of turning left is synonymous with the driver having the intention to turn left. Several definitions of human error are, in fact, framed around the concept of intention and the difference between intentional and unintentional acts. However, not all approaches do so, and more seriously an overreliance on the common language meaning of intention can be misleading.

To illustrate the point, I will use two well-known definitions of slips and mistakes that rely on the concept of intention. Norman (1983) provides a very concise characterisation of slips and mistakes:

> If the intention is not appropriate, this is a mistake. If the action is not what was intended, this is a slip.

This statement contains ambiguities in the use of the term intention. Since it aims at defining human error in the context of real work tasks (which are typically characterised by multiple goals, interdependences between goals, time constraints, sub-goals, preconditions, execution conditions, etc.), Norman's meaning of intention is 'plan' (a rule, both actions and goals) in the case of mistake and 'expected outcome of the plan' (actions implied by the plan) in the case of slip. The intention of Norman's definition of slip is clearly the 'goal' of the actor, the expected outcome of his/her plan: the operator intended to push a button but accidentally pushed another one. In the case of a mistake, the intention cannot strictly be the goal of the actor otherwise it would be the common definition of a violation (including acts of sabotage, suicide attempts, etc.). Instead, it is the plan (goals and actions) that is inappropriate to achieve the intention (overall goal), the plan that is inconsistent. A plan P is inconsistent when:

(1) It does not imply a specific execution E to take place: (P does not imply E)
    or
(2) When the execution E that it correctly implies is not adequate to achieve the overall goal (OG) of the plan: (P implies E but E does not imply OG).

Generally, the selection of a wrong goal – for example, due to lack of skills – is not considered a violation. This is exactly the problem with the meaning of intention: the wrong goal selected here is properly a sub-goal (SG), that is to say, it is a means to achieve an overall goal (e.g. secure the system). In terms of our definition of inconsistent plan, in this case of complex tasks, a plan is inconsistent when:

(1b) It does not imply a specific execution E to take place: (P does not imply E *or* E cannot achieve SG)
    or
(2b) When the execution E that it correctly implies is not adequate to achieve the overall goal OG of the plan: (P implies E, and E implies SG, but SG does not imply OG).

So, again, 'choice of a wrong goal' is a mistake, not as an inappropriate intention but as inadequateness of the means (the sub-goal) to achieve the top goal (as in (2b) of the above definition). Clearly, Norman did not mean 'inappropriate top-goal' in the definition of mistake, but rather 'inappropriate plan' in the sense of inappropriate means for the overall goal. The point is, however, that the different meanings of intention and the different levels of analysis of the task in the definition are not made explicit. As a consequence, Norman's statement switches between intentions as expected outcomes (in the case of slips) and intentions as plans (mistakes) (or between simple and complex tasks).

Another example of the difficulties of working with intentions is provided by Reason's (1990) working definition of human error:

> Error will be taken as a generic term to encompass all those occasions in which a planned sequence of mental or physical activities fails to achieve its intended outcome, and when these failures cannot be attributed to the intervention of same chance agency.

This definition is ambiguous because it does not specify whether the outcome is the object of the intention of the actor or of someone else. When mistakes are characterised as 'failures of intended actions to achieve their desired consequences' we would not like the 'desired consequences' to be exclusively those of the agent, otherwise we would have to rely (typically) on the actor's subjective experience of having made an error in order to characterise the action as a mistake. To recognise their own mistakes as well as to correct their plans before undesired consequences occur is a very useful skill of human beings, but it is not a general condition on which to base a definition of human error. This ambiguity is not resolved in Reason's proposed working definition for mistakes:

> Mistakes may be defined as deficiencies or failures in the judgmental and/or inferential processes involved in the selection of an objective or in the specification of the means to achieve it, irrespective of whether or not the actions directed by this decision-scheme run according to plan.

Leaving aside the issue of referring to the judgemental and inferential processes as the inferred causes of mistakes, a mistake is defined as a failure in (1) the selection of an objective or (2) the specification of the means to achieve it. The first statement of the disjunction is similar to Norman's definition of mistake with a potential ambiguity between objective as overall goal and objective as sub-goal, and hence between mistakes and violations. When the objective is read as the sub-goal, the definition is on the whole coherent with our characterisation of inconsistent plan for complex tasks. In other words, if we assume objective to mean sub-goal, i.e. we are not concerned with malevolent violations (as Reason was not, in the chapter in which he put forward the definition), the ambiguity will be removed. Still, the intention here seems to be exclusively the one of the agent and we have seen that in many cases failures and unsafe acts are defined in relation to other criteria.

1.5 Violations

When we take the issue of violations seriously into account the story becomes even more complicated. So far, we have assumed a simple sense of violation as 'malevolent violation', that is, a deliberate choice of goals contrary to the system's goals, as in acts of sabotage and vandalism. A violation, then, is very easily recognisable as we assume that the individual is capable of choosing between well-understood and unambiguous system's goals. In reality, this is not straightforward. The individual can disregard but also misunderstand the prescribed task for a variety of reasons: because of lack of knowledge, because the goals are poorly defined, because the system contains conflicting goals and principles, because the conditions of execution do not make it possible to perform the task in all situations, etc.

Following a scheme proposed by Leplat (1993), in all such cases there is divergence between the prescribed task or 'task for the expert' and the redefined task or 'task for the subject'. When the subject knows the prescribed task, but for some reasons does not want to execute it we would normally call it a violation. This is true from the point of view the agent's intentions, and is the common interpretation of a violation. In this view some degree of intentionality or deliberation must be present to qualify a divergence between the prescribed task and redefined task as a violation.

Furthermore, for Leplat (1993):
> The error for the expert defines the divergence between what it is expected by a subject (result, procedure ...) and what it is really done. The error for the subject is the divergence between what he/she wanted to do and what he/she thinks he/she has actually done.

In this scheme, for the expert the violation would be the divergence between what the 'expert' expects, i.e. the prescribed task, and what the subject intends to do, the redefined task, when this divergence is not due to a misunderstanding of the task, in which case we would rather talk of a mistake (for the expert). The subject would generally be aware of the divergence (i.e. that there is some different expectation of his/her behaviour), at least at some level of consciousness. In other terms, he/she knows the prescribed task but cannot or does not want to follow it in some circumstances.

In Leplat's scheme the 'task prescribed' does not only include previously specified procedures, rules and instructions but it also includes vaguely specified 'missions', the principles to be considered in the evaluation of a task, as well as the conditions of execution. In this sense the term violation would not apply to cases of deliberate choices of goals contrary to rules and regulations, but in which the agent's intention was not in contrast with the system's overall goals (e.g. safety). This is a consequence of the fact that the 'task for the expert' is an ideal and (by definition) correct prescription, while the actual prescriptions embedded into work procedures, rules and orders might sometimes be inadequate or neutral towards the realisation of the system's overall goals.

If we shift to an external point of view, one could call violations also those cases where the subject did not know or did not deliberately choose not to follow the prescribed task (actual or ideal). One can find in the literature examples of rule violations that are attributed to lack of training or understanding which clearly point to the fact that the subject did not know he/she was not following the prescribed task (Reason's 'erroneous violations', which could in fact be considered 'mistakes'). Also, it is common in the violation literature to talk of routine violations, behaviours contrary to rules and regulations which have become the norm, that is, executed automatically and without deliberation.

Once an external point of view is taken, the realm of application of the concept violation extends to include all behaviours that diverge from procedures, rules, instructions, and 'missions', as well as from the principles to be considered in the evaluation of a task, together with their conditions of execution (possibly everything not due to impairment, as in legal terms). In other terms, the conceptual distinction between errors and violation is far from being clear-cut. What is certain is that in defining and evaluating unsafe human actions we must be aware of the differences in relying on internal versus external points of views as well as of the consequences of assuming actual versus ideal standards of correct executions. Table 1.2 shows how a 'phenomenology' of violations can be obtained by considering these two dimensions.

Table 1.2 Phenomenology of violations.

	Point of view	
Standard of correct execution	**Internal: looking at the deliberate choice of goal**	**External: looking at actual behaviour**
Ideal: objectives and principles	**Malviations**: Malevolent and irresponsible intentions, incorporating certain or likely negative outcome. Examples: – Sabotage, vandalism, etc. – Situational, exceptional violations.	**Mistakes**: Misunderstanding or ignorance of system's objectives, principles and conditions of application. Example: – Reason's 'erroneous violation'.
Actual: existing rules and procedures	**Workarounds**: Goal conflicts, i.e. choice of *systems or personal* objectives and principles which conflict with *known* existing rules and procedures. Also, onflicts among rules, with both positive and negative outcomes. Examples are: – Non-formalised best practices and recoveries – Non-harmful short-cuts, strategies, etc. – Situational violations, case adaptation of inapplicable or conflicting rules – Exceptional and optimising violations.	**Side-tracks**: Ignorance of existing rules and procedures, but also non-deliberate behaviour contrary to rules: – Behaviour dictated by *system's* objectives and principles which conflicts with *ignored* existing rules and procedures; generally associated with positive outcome – Routine violations.

1.6 Intentions

It should now be clear that the problems and ambiguities discussed in relation to Norman's and Reason's definitions, and the various interpretations of the concept violation, rotate around the meaning of intention and intentional behaviour, and between the difference between the intentions of the expert, i.e. system designer, management or society, and the intention of the subject. It should also be stressed, however, that Reason's and Norman's definitions are working definitions, and as such their appropriateness lies in their utility. The problems discussed stem from the multiple meanings of the concept intention and the fact that it is present, as standard of correct performance, in the definition of all types of unsafe acts, whether they are called errors or violations.

The term intention can have three different meanings that can be outlined by recalling the history of the philosophical use of the concept. In Latin, *intentio* originally overall the same meaning as 'concept' but later was used by medieval philosophers, first of all by Thomas Aquinas (1225–1274), to indicate both the reference of the concept (an objective entity) and the act to refer. The concept was reintroduced in the 19th century by Austrian philosopher and psychologist Franz Brentano (1838–1917) to define all psychological phenomena, as opposed to the physical ones. For Brentano, all psychological events were intentional in the sense that they are directed toward some object; they relate to some content. In addition, all psychic acts, insofar as they are intentional, are completely present to the consciousness; they can be entirely known.

These aspects of the concept of intention are still present in the common use of the word, as we have seen in the definitions above. We have seen that the concept intention is thought to have the following properties:

(1) It is the expected outcome of an activity, i.e. the goal (parallel to the referred object).
(2) It is the outcome *and* the actions to achieve it, i.e. the plan (parallel to Aquinas' *intentio*).
(3) It is a mental phenomenon present to the actor's consciousness, e.g. the deliberate violation (as in Brentano's psychic act).

When the three aspects of the concept intention are clearly recognised it becomes easier to understand the concepts of 'human error' and 'unsafe act' as well as to interpret the definitions that make use of them. Further, it would probably be less misleading to think of mistakes in terms of the second meaning of intention given above, i.e. as inappropriate plans or inconsistent plans of actions. Keeping in mind the previous discussion, we may define three classes of unsafe acts *from the actors' point of view* in their most basic form: (a) slips as wrong executions; (b) mistakes as wrong plans of actions; and (c) malevolent violations as wrong intentions (as top goals).

It is also true that some characterisations of human error do not refer to agents' intentions at all, as for instance in the engineering tradition. It is, however, necessary to refer to intentions, volitions or reasons in order to provide psychologically tenable treatments of human errors and violations. It is also natural to refer to intentions when error is defined from the point of view of the agent. However, it must be pointed out again that when analysing actions with unwanted consequences, the intentions are not always those of the agents but the standard of correctness for the actions can be external (e.g. procedures, expectations of the organisations, etc.) and may or may not coincide with those of the agents.

1.7 Violations as errors

The everyday meaning of the term 'violation', as well as the insistence on the agent's point of view and intentions, suggest that there is something peculiar about the concept, in particular that there is a radical difference from the concept of error. The previous discussion has shown that this is the case only if we restrict the use of the term 'violation' to the class of 'malevolent violations', i.e. deliberate choices of goals contrary to system's goals. Furthermore, as Dougherty (1995) has convincingly shown, violations can be reduced within a standard taxonomy of error, provided that the model underling the taxonomy is extended to include the domain of judgements under risk.

Dougherty (1995) defines violations as *'behaviour that results when a person assesses that the formal expected response is inadequate'*, assuming that the primary goals of the person are not contrary to the system's objectives and principles. This definition would then fall in the 'workarounds' box of the phenomenology of violations presented above (Table 1.2), that is, it is an agent's point of view definition restricted to choices made against known rules and regulations.

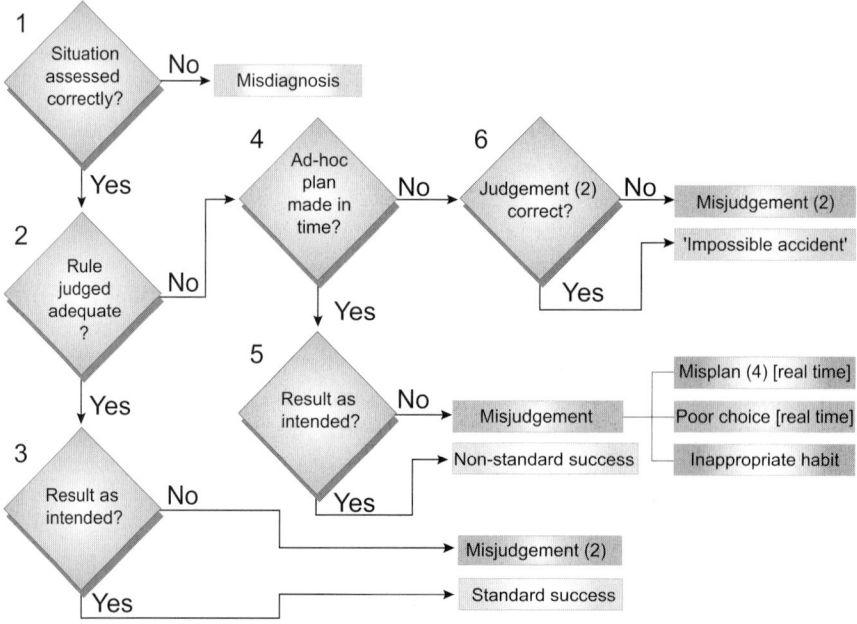

Figure 1.1 'Violations as errors' flow diagrams.

Figure 1.1 depicts a flow diagram for reducing rule-related behaviour to successes, misdiagnosis or misjudgements.

Dougherty's (1995) proposed version of the definition of a violation implies that the assessment of a rule's applicability to a situation is a judgment that might prove to be wrong, i.e. have unintended or unfortunate results. That is, a violation might be a misjudgement and may clearly include emotional and motivational aspects. Misjudgements are sub-divided into misplans and poor choices. The difference between a mis-plan and poor choice is one of cognitive content, the two being the extremes of a scale where also snap judgements, decisions and deliberations could find place. Violations would then be indicators of difficult decisions, and after the fact, of bad judgments.

Although on closer scrutiny the substantive difference between the concepts of error and violation seems to have vanished (at least when looking at the deliberate choice of goals), one could maintain a practical distinction between errors and violations regarding two different avenues to complex systems' safety improvement (Table 1.3). Errors are the cognitive category associated with the disciplines of ergonomics and human reliability. Here, the focus is on the individual. Violation refers to concepts such as individual motivations and attitudes, which in turn refer to the social aspects of behaviour, and which normally fall within the areas of organisational research and safety culture. The focus is then the organisation. The distinction will ultimately be in the different instruments which are applied in individual versus organisational research.

Table 1.3 Contrasting cognitive and psycho-social approaches.

Cognitive approaches	Psycho-social approaches
Human error occurs when mental demands exceed capacity – Applies well to automatic or semi-automatic cognition and laboratory tasks	Human error as a consequence of difficult decisions – Decisions under goals and/or rules conflict, stochastic uncertainties – Often real-time, 'hot cognition'
Capacities and demands are context dependent – Performance Shaping Factors	The context relates to high-level decisions – Perceived management pressure rather than task complexity
Fit with situations where technical systems require human performance at its limits – Combat aircraft – Individual micro-tasks in Nuclear power plant (NPP) control room	Fit with human performance where non-cognitive factors are stronger – Motives, group pressure, production pressure – Normally there is sufficient time, attention, etc.

1.8 Violations as 'non-compliance'

The term violation is inextricably interlinked with the concepts of intention and deliberation, and thereby with intentional acts. Yet, the field of road transportation research provides ready examples of violations as unintentional acts: a car driver passes a red traffic light unintentionally because he/she did not notice it. In this field, exactly the same event can be explained by a completely intentional act, as when a driver decides not to stop at the red light because he/she is in a hurry.

In the human error literature there is a prevalence of defining violations in terms of intentional behaviour. Reason (1990, p. 195), for example, defines violations as 'deliberate – but not necessarily reprehensible – deviations from those practices deemed necessary (by designers, managers and regulatory agencies) to maintain the safe operation of a potentially hazardous system'. Similarly, Mason (1997, p. 288) describes violations as 'deliberate deviations from the rules, procedures, instructions or regulations introduced for the safe or efficient operation and maintenance of the equipment'. However, in the same tradition, Collier (2000 pp. 3–4) considers violations to 'span the full range of intention [as they] include acts that are totally habitual and "unconscious", as well as fully deliberate acts'.

Collier's position is in fact more consistent with human factors practice. Both Reason and Mason accept a classification of violations that includes 'routine violations', a behaviour contrary to the rules that has become the norm, and which is executed in a 'automatic and unconscious' (Mason, 1997) manner. Reason (Reason et al., 1990), in the road traffic context, labels the one he reputes to be the most interesting class of violations as 'erroneous or unintended violations', a type of activity that is deliberate but which does not have the prior intention to cause injury or damage.

Beside malevolent violations that are intended to cause harm or damage, i.e. acts of sabotage and vandalism, there is a potential overlap between what would be defined as error and what as violation. This overlapping set can, in a context of operation regimented by rules and regulations, be better referred to as 'non-compliance' or 'non-adherence' behaviour. Such expressions have the advantages of:

1) Suspending the judgement on the righteousness of the intentions or the goodness of the rules and regulations not followed.

2) Restricting the realm of application to existing rules and regulations as the 'standard of correctness'.

3) Being independent from the result of the behaviour (i.e. non-compliance can lead to both success and failure).

In such a way, violating behaviour would be such only if the context of rules and procedures determined it to be so. As Reason (1990, p. 195) puts it, 'violations can be described only in relation to a social context in which behaviour is governed by operating procedures, codes of practice, rules and the like'.

A product of operationalising violations as non-compliances is that the link between behaviour and systems failure is totally mediated by the rule systems. Only if a rule system is 'error proof' will successful compliance guarantee the absence of failures. In other words, the concept of non-compliance is not directly applicable to risk analysis. Instead, it is appropriate for evaluating the status of the safety rule system of an organisation, by means of mapping the level of non-compliance and investigating the reason for this form of behaviour (see for example Massaiu, 2006).

1.9 A general definition of human error

The previous discussion has outlined the fundamental dimensions necessary in defining human error: the goal or intention, as the standard of correctness, and the action to be evaluated. We have concluded on the importance of restricting the meaning of intention in order to differentiate between plans and goals. Connected to this is the level of application of the definitions: primitive tasks versus complex ones provides yet another way to confound between intentions as plan, intentions as overall goal, and intentions as sub-goals. Keeping in mind these distinctions, I define human error in the following way:

> Human error is the failure to reach an intended goal, the divergence of a fact from a standard.

This definition is able to include all of Reason's types of unsafe act (slips, lapses, mistakes, and violations), by way of selecting the appropriate goals and intentions. The standard of correctness can be internal to the person (the person's intention or expected consequences of his/her action), or external (the expectations that other persons or organisations place on the agent). When reference is made to internal standards of correctness it is not required that we have to rely on the person's own experience of having made an error. This experience can be valuable or not, depending on the circumstances, but it is not necessary. There are, in fact, external or public criteria that, through inference, allow for the ascription of intentions to the agents, in the way that is typically done in cognitive psychology. By reference to goal structures, volitions and intentions, it has been possible, for example, to distinguish between mistakes and slips, that is, between actions that follow an inadequate plan and actions that follow an adequate one but which fail to reach their goal.

It may be questioned whether the goal has to be present in the agent's consciousness, that is, as an explicit goal, or whether it can be sub-conscious or unconscious, that is, as an implicit goal, as in the case of lower level cognitive tasks such as motor skills. The point is clearly related to the issue of routine violations, that is, to the degree of deliberation of the action being evaluated. The answer is that since the process of error attribution is a normative one that normally is not performed by the subject(s) who committed the actions at issue, the difference is not important, as external criteria or internal attributions are employed as standards of correctness. It becomes important in terms of error psychology, where the internal mechanisms of error are the issue of study (in this respect see Reason, 1990, pp. 5–8).

Finally, it should be noted that this definition of human error would correspond to a definition of error in general were it not for the nature of the goals. It is the cognitive and intentional nature of the goals that makes these errors 'human'.

1.10 Practical aims of error analysis

All theories and techniques that treat the issue of human error necessarily refer to some combination of the following three causal factors: 1) person-related/psychological; 2) environmental-external; and 3) task characteristics. Differences in characterisation, importance and interactions assigned to these three elements result in different theories, models or approaches to human error, as we will see in Section 2. The relative importance of the causal factors present in an explanation of human error is, moreover, dependent on the main research question. It is not difficult to identify the three most common issues in retrospective and perspective error analysis:
1. The event is the cause of unwanted consequences
2. The actions are caused by some internal and external factors
3. The actor is responsible for the unwanted consequences.

The three issues are traditionally associated with different disciplines. The first is exemplified by the engineering approach. The traditional engineering approach (before the World War II) identified incident causes into 'unsafe acts' and 'unsafe conditions', that is, attributed the cause of systems' failures to either humans or equipment. Accident prevention manuals at that time attributed 80% of incidents to humans and 20% to equipment (Heinrich, 1931). The human and technical causes were seen as independent of each other and the prevention strategy to be the modification of either one.

More recent engineering approaches, known as human reliability assessment, still start from the distinction between human and technical failures but have enriched the analysis. After the human or technical sources of system failures are identified, the analysis can go further in identifying the components' subsystems or operators' functions that failed. The decomposition will stop at the failed sub-components or human functions for which adequate failure probabilities data are available. For instance, in the case of an operator that failed to start the auxiliary feedwater system a fault tree diagram will be produced where the operator failure is represented in terms of combinations of elementary task functions necessary to accomplish the task, e.g. read an analog meter, diagnose an abnormal event within 10 minutes. What is still common to the old approach is that human failures are defined in terms of unfulfilled operator functions, or unperformed assigned tasks, and not from the point of view of the subjects.

The second issue is the core of the discipline of error psychology. The interest here is in the psychological causes of the action that failed (independent of it having negative consequences on a particular occasion). Clearly, this perspective complements the previous one by providing the theoretical basis for a fault tree specification and (at least in principle) failure data. Error psychology investigates the psychological mechanisms that control cognitive activities and identify internal mechanisms, psychological functions or global performance control modes together with tasks conditions as causes of errors.

The perspective represented by the third statement is typically a juridical or moral one. It aims at establishing the degree of involvement and the margins of choice of the agent in the causal process that led to the unwanted consequences. The themes of intention, comprehension and autonomy are central in answering this question. This issue is related to accident investigation, although most techniques limit their scope to the multiple causes of an accident and leave the issue of personal responsibility to the prosecutors. As practical enterprises, accident investigation techniques use methods and models from the two previous approaches.

It should be stressed that the issue of responsibility has a bearing on the topic of human error at work well beyond an accident investigation perspective, as the degree of responsibility associated with a task influences the behaviour of the agent in purely cognitive and behavioural terms (Skitka, Mosier, & Burdick, 1999; Skitka, Mosier, & Burdick, 2000). Thus, the issue of responsibility can itself be a causal factor of accidents and should be considered in the design process (e.g. function allocation, support systems, error tolerance).

It is clear from this discussion that different approaches and research questions focus on different aspects of the causal explanation of human errors, although they all necessarily include, at least implicitly, reference to the three levels mentioned before: psychological, environmental and task. Yet, the reference to the core factor permits differentiating the different approaches of human error modelling and classification.

1.11 Errors, accidents and safety

The classical paradigm of safety science maintains that, in order to achieve safety, hardware failures and human errors must be reduced or eliminated. The study of accidents and incidents is natural field in which to learn about errors, since the analysis of past events makes it possible to identify systems' failures, discover their causes, and in this way generates general knowledge. This direction started already at the beginning of the 19th century and were directed by the following assumptions: (1) that there are two paths towards incidents, that is, technology failures and human errors; and (2) that the two are quite independent from each other. These two assumptions remained the hallmark of safety science up until the 1980s, and their influence is still strong today (as is readily apparent from the media treatment of technological accidents, which typically ask whether the cause of the accident was technical failure or 'human error').

As technological progress in the 20th century advanced faster than human factors science, this traditional view on safety, which maintained two independent causes of accidents, ended up placing considerable emphasis upon the negative influence of the human element, and in particular of 'front line' operators of the systems: pilots, air traffic managers, ships' officers, control room crews, anaesthetists, and so forth. The major system safety challenge soon became the reduction of the potential for human errors as the dominating cause of accidents. A first solution was envisaged in designing the human out of the systems by introducing mechanisation and automation. When this was not possible, and hence the human element had to be left a place, the inclination was to apply to the human the same theories and methods as applied to the hardware elements of the system. An example of this propensity is the Fitts' list, which compares humans and automatic machines against the type of task they can perform, as a means to allocate functions in a system. As we will see later (Section 2.1), such early approaches did not contribute much to the reduction of accidents nor to the understanding of the human role in systems' safety. They lent ideological support to the '80:20 rule', an unproven assumption that stated that 80% of accidents were human caused and 20% equipment caused, to the extent that this became common wisdom in accident prevention manuals of the time.

The reason why these early approaches did not advance knowledge on risk and safety was that they had serious methodological flaws. Incident analysis, framed into the human-machines dichotomy, did not allow for finding general patterns in particular incidents. As incidents are typically the result of unique mixtures of factors, the reliance on a simplified causal model made it impossible to identify the real determinants of accidents, to the point that even the distinction between causes and effects became arbitrary. In fact, these early attempts lumped together incidents independently of their characteristics and especially independently of the human contribution to the events. The role of the individual in the accidents was not really modelled, apart from the psychophysical characteristics of the victims.

1.12   Modern safety science

Safety research thus tried to understand why incidents occurred as well as envisage remedies for accident prevention. However, the study of human error as a specific topic only came to the forefront of industrial research late in the 20th century and as a consequence of large-scale accidents, such as the Tenerife aircraft collision, the nuclear power incidents at Three Mile Island and Chernobyl, and the Space Shuttle Columbia's failed re-entry, to only mention a few.

The old dichotomy between technology failures and human errors was replaced by system thinking. The modern approach considers safety as the result of the interplay between individuals, technology and organisations, a perspective that in Scandinavia is typically referred as man-technology-organisation (MTO). This renewed system safety science recognises the inadequacy of treating humans with the same tools and methods as used for the hardware elements, and special emphasis is placed on the disciplines of human factors, applied psychology and organisational research. The leading findings of about 20 years of cross-disciplinary research on the role of human error in systems' safety have modified the intuitive assumptions normally associated with the relation between errors and accidents. They can be summarised as follows:

1. Human errors have to be viewed in a system perspective in order to assess their contribution to safety. Individual errors can and do occur without resulting in accidents: most human-machine systems incorporate barrier functions or safety nets that bring the system back to safe operating conditions in cases of

deviations caused by initial failures. Amalberti (2001) provides a quantitative estimate of one human error out of 1000 that have unacceptable severe consequences. It is now accepted knowledge that accidents in ultra-safe production and transport systems (i.e. systems with less than one accident per 100,000 events) are usually the result of unforeseen combinations of errors happening at different levels of the man-technology-organisation complex. The ideas of defence-in-depth (Reason's Swiss cheese model) and high reliability organisations (Rochlin, 1993) were developed in this context.

2. Human errors cannot and should not be eliminated completely. As it became clear following the first international conferences on the issue (Senders & Moray, 1991, Rasmussen et. al., 1987), human error should not be treated in exactly the same terms as technical failures. In the first place, it was noticed that errors are an essential component of learning, and that they even seem to have positive roles, e.g. in terms of promoting creativity, exploration and adaptation. Even more importantly, although humans often produce errors that result in accidents, they more often perform correctly and, in particular, are capable of detecting and recovering both system errors and their own errors. Detection and recovery of error might even be considered as better indicators of performance than error production.

3. Individuals recover the majority of their own errors before they result in incidents. This means that error control is part of the broader performance control, the cognitive regulation of performance where operators dynamically optimise performance objectives and costs. Cognitive control includes activities such as: awareness of performance goals and difficulty at the required level; style of control used (conscious or automatic); choice of mechanism to detect and recover errors; and tolerance towards produced errors. The ideas of cognitive control and recovery potential have resulted in two classes of approaches. The first class is known as error management, error handling or simply error recovery. System safety is pursued without concentrating on errors per se but on the generation and propagation of system hazards and in the processes of preventing them in resulting into accidents. These approaches, that in the literature come under the names of error management (Bove, 2002), treat management (Helmreich, Klinect, & Wilhelm, 1999), and control of danger (Hale & Glendon, 1987), provide models and classification of human error which are different from those that concentrate on human error production mechanisms. The second class studies the cognitive control of global performance and individuates, for example, cognitive control modes (Hollnagel, 1993) or the meta-knowledge and confidence that ground cognitive risk control (Amalberti, 1992).

1.13 Classification

A classification of error is a structured way of reducing the multiplicity of error manifestations to a smaller set of fundamental manifestations or to a set of causal mechanisms. In principle, error classifications or taxonomies are not different from those found in the natural sciences. In practice, error taxonomies lack the internal systematic order of the natural taxonomies which are organised around few and simple principles. The problem is that in the field of human error there are neither agreed definitions of what constitutes the manifestations that are to be organised, nor simple causal relationships between causes and between causes and manifestations.

The causal explanation of behaviour (and thereby error) is the base of a classification system. Without a causal model a classification scheme is arbitrary since it is the underlying model that determines how the scheme is organised, what is cause and what is manifestation, how the terms are to be interpreted and applied, and what combinations are meaningful. As different causal models exist to describe the complexity of human behaviour, so too are there differences in the description of human errors between and among taxonomies. In general, there are two levels of description of human error. The basic level of description is the overt behaviour or manifestations of errors as discussed above (for example, omission and commission, wrong timing, too much force). Classifications that include characteristics of the individual, of the internal psychological mechanisms and of the external environment refer to the causes of behaviour and not only to manifestations. Such causes can be observable, e.g. feature of the situation such as glare, noise, equipment, availability of procedures, years of service, or they may be theoretical constructs hypothesised to explain cognitive processes, e.g. decision, diagnosing, capacity limitations, observation, etc. Errors as causes can be divided in terms of such internal functions, e.g. errors of detection, decisions errors, or they can be related to the features of the situation, e.g. stress-related error, poor illumination and glare.

Beside the causal model adopted, error classification can be organised around the principle of risk management or control of danger mentioned before. In this case the classification and modelling will not be limited to errors' causes and manifestations but will include the wider process of successful and unsuccessful performance. This process is centred on the way errors and hazards are handled more than on the way errors come about. It should be noticed that error producing and error management approaches are not theoretically contrasting views but rather the difference is in the task performance level used as unit of analysis. The point can be illustrated by contrasting risk management in air traffic control with human reliability analysis in the nuclear sector. The latter has the main goal of quantifying the reliability of man-machine systems, typically a nuclear power plant. System experts write down a PRA/PSA (Probability Risk/Safety Assessment) event tree model, which is a logical representation of how a set of disturbances (initiating event) can develop into a serious negative outcome (e.g. core damage). Operators' activities are usually represented as recovery behaviours that need to be assigned a failure/success probability in the same way as all other failures represented in the event tree model. Similar logical models, called fault trees, are often used to calculate failure probabilities. In the case of human failures the required recovery behaviours are typically decomposed into the logical combinations of operations and cognitive activities necessary for their success. Human error probabilities for the undecomposed events are obtained from published sources or estimated by the experts and are adjusted for the effects of contextual factors present during the performance (performance shaping factors). The example shows that HRA (human reliability analysis) models only the human error production phase while error management is incorporated in the system analysis, the PRA, which properly provides the starting point of the HRA. System experts thus perform the task of modelling system- and risk-scenario dynamics in the PRA before the HRA is performed. This rigidity in the modelling of a dynamic system has been repeatedly criticised (Hollnagel, 1998) but is dependent in part on the, at least assumed, predictability of the process of nuclear power production and in part on the quantification requirements.

In human machine systems where risk scenarios and dynamics are less predictable and the focus is not on risk quantification but rather on risk identification and reduction, the phases of error production and error management are typically analysed as parts of a single process. This is generally the case in aviation, air traffic control and road traffic, and is the hallmark of incident investigation. Here it is customary to analyse large performance segments or series of events where (possibly) different actors perform many activities, and where errors are committed, recovered or exacerbated in the risk management process. I will return later to risk management models and classifications, suffice it to say here that, besides the focus on whole performance success or failure, these approaches emphasise the positive side of human performance and the active and anticipating role of the operators.

This discussion also shows that there is a strong relationship between the theoretical approach, the practical purpose and the domain of application, which determines the level of description and the shape of the classification systems. If we concentrate on the purposes for classifying human error we can specify four main classes:

1. Incident investigation. To identify and classify what types of error have occurred when investigating specific incidents (by interviewing people, analysing logs and voice recordings, etc.).
2. Incident analysis. To classify what types of error have occurred on the basis of incident reports. This will typically involve the collection of human error data to detect trends over time and differences in recorded error types between different systems and areas.
3. Error identification. To identify errors that may affect present and future systems. This is termed human error identification (HEI).
4. Error quantification. To use existing data and identified human errors for predictive quantification, i.e. determining how likely certain errors will be. Human error quantification can be used for safety and/or risk assessment purposes.

Incident investigation and incident analysis are retrospective activities where the classification system will help in explaining an event that has already happened. Most classification schemes are developed for retrospective analysis. Error identification and error quantification are predictive analyses, where the interest is in events that can happen. Predictive analysis has been the concern of system designers and reliability practitioners. Although the explanation of past events and the prediction of future ones are the basic features of any scientific theorising, the exchange of methods and of classification schemes between the two directions has been rather limited. This is due to lack of comprehensive theories of human behaviour and the consequent need to delimit the scope of the analysis to the prevailing interest. Another point of difference between prediction and retrospection is that while reliability studies have centred classification at the observable level of behaviour (e.g. omission and commission), incident investigators and system designers have favoured descriptions at a deeper causal level.

2 Human error and models of behaviour

In the process of error attribution, or the equivalent, namely evaluation of normative statements, it is essential to specify the standards of correctness adopted as well as the model of human performance that controls the application of the standards to the conditions of execution under investigation. In the words of Woods & Cook (2003), 'the standard chosen is a kind of model of what it means to practice before outcome is known. A scientific analysis of human performance makes those models explicit and debatable'. It is in this spirit that in this section I will describe the main models of human performance that have been used in the study of human error.

Behavioural theories have always used models and metaphors to explain the complexity of human mind and behaviour. A number of these have been borrowed from the prevailing scientific and technical paradigms: mechanics and steam power in the 19th century, animal learning and telephony in the early 20th century, computers after the World War II, and, more recently, cybernetics and artificial intelligence. The dominant psychological schools of the early 20th century were psychoanalysis and animal learning. The former exerted its influence in therapy and the media, while the latter dominated academic psychology, particularly in the United States. In this respect, the most influential psychologist was John B. Watson with his 'behaviourist manifesto' of 1913 (Watson, 1913), where he banned the mentalist tradition, that is, the discourse on mental concepts such as intention, volition, and particularly consciousness and introspection. Parallel to the animal learning and behaviourist psychology was the controversy over heredity and environment – nature versus nurture – that framed the investigations into industrial accidents. Those who believed in the centrality of heredity developed theories that explained behaviour in terms of observable individual characteristics. In criminology theories were developed that classified criminal types by physiognomy. Similarly, in industrial accident investigations individual characteristics, such as sensory capacity, speed reaction and personality, were looked upon as determinants of the likelihood of a person being involved in accidents. This early approach to describe human behaviour at work went under the name of accident proneness theory, and is the first model of behaviour that I will describe in this section.

2.1 Accident proneness model

The accident proneness model was developed in Great Britain at the end of the 19th century and the beginning of the 20th to explain the increased accident rate in industrial production. The theoretical context was the heredity versus environment controversy, which in turn was rooted in Darwin's evolutionary theory. Two explanations were advanced to explain the increasing rate of industrial accident: the first stressed the importance of the environment, that is, the growing speed of production and the more demanding work tasks; the second regarded the individual differences as being more important and was historically concerned with the consequences of drafting regular workers for World War I and the employment of (assumedly) less competent young men and women. The dispute was never resolved and was probably irresolvable in the way it was posed. The premise of debate was, in fact, that the two explanations were independent of each other to the extent that the individual characteristics would make some persons more dangerous independently of the technical environment. As a matter of fact, the environmental perspective succeeded in guiding health and safety regulations, as documented by accident prevention manuals of the time. The accident proneness model, on the other hand, guided accident investigations and research, and became popular among insurance companies.

The accident proneness model claimed that individual differences made some persons more likely to be involved in accidents. Hence, it researched individual differences of sensory (e.g. visual capacity), psycho-physical (e.g. speed of reaction), and psychological (e.g. personality) nature. The results of the research were generally poor and no psychological classification of accidents was produced. Hale and Glendon (1987) summarise the shortcomings of the accident proneness research as follows:

(1) The proneness could be 'proved' only after the incident, hence statistics emphasised the characteristics of the individuals rather than those of the accident.
(2) Accidents were grouped together for statistical analysis regardless of their characteristics and the real involvement of the victim in the accident causation.
(3) The preventive actions proposed by the model were (a) excluding some individuals from performing dangerous work or (b) modifying mutable traits by training, counselling and motivation.
(4) Different groups of individuals, however defined, could and were found to have higher accident rates but psychological characteristics were unable to explain more than 20% of the variance in accident rate.
(5) The theory offered the opportunity for blaming the victims for the accidents, thereby shifting responsibility away from the employers.

The failure of the accident proneness model to find a valid set of explanatory individual factors, which could be used for accident prevention, discredited not only the model but also any psychological attempts to provide a practical basis for system safety improvement. Designer and engineers, lacking the basis for differentiating between the normal and the accident prone, assumed the worst-case scenario, that is, they assumed all humans to be unreliable and sought system safety by reducing the human role and increasing automation. This conviction was further reinforced by the indirect support provided by the accident proneness model to the 80:20 rule, which stated 80% of the accident to be human caused and 20% equipment caused: the reduction of the 20% of technical causes became a measurable objective for safety research.

2.2 Traditional human factors and engineering models

Engineering approaches to system safety have maintained the dichotomy of human versus technical failures. There are, however, two interlinked perspectives regarding how to treat human failure. The first perspective considers human failures in all stages of system life cycle – specification, design, manufacturing, installation, maintenance, modification and, not least, operation – as systematic error, that is, error with identifiable and modifiable causes, which in essence is a non-probabilistic phenomenon. The alternative approach considers human failures, particularly human errors, during operation to be random, at least at the elementary task level, and hence to be quantifiable as probabilities of failure. This approach is the essence of techniques of human reliability analysis (HRA) that are part of systems' probability risk assessment (PRA), which I will discuss later.

2.2.1 Traditional human factors

Human factors, or human factors engineering, can be defined as applied research on the physical and mental characteristics, capabilities, limitations, and propensities of people at workplaces and the use of this information to design and evaluate the work environment in order to increase efficiency, comfort and safety (Kelly, 1999). Human factors became firmly established as a separate discipline during World War II as a consequence of the proliferation of highly complex systems (particularly aviation systems) that stretched human capacities to their limits. Human factors practices and standards have since become a major consideration in many design areas, particularly those in which the human/system interface is critical to overall system safety. Human factors research and recommendations address such issues as automation and control, military system design and nuclear power plant regulation and evaluation, as well as consumer usability issues such as the layout of automobile dashboards. Human factors research maintains that most active monitoring and intervention by operators in complex systems involves cognitive (mental) functioning. Typical study issues are fatigue, memory, attention, situation awareness, workload, cooperation, training, manpower, crew management, and decision-making.

Insofar as the discipline of human factors is concerned with the problem of system design and production of standards and regulations, the focus is on the global and qualitative aspects of human performance. Human error is treated indirectly, based on the assumption that improved system design solutions will aid human activities and hence reduce the occurrence of errors. The definitions of human error are typically framed from the point of view of the subject with reference to cognitive processes and the context of execution. That is, *human error is viewed as degraded performance determined by a complex set of causal factors.*

The view is neatly exemplified by a joint European effort to harmonise the safety standards of railway signalling through the European Committee for Electrotechnical Standardization, CENELEC. The CENELEC standards assume that safety relies both on adequate measures to prevent or tolerate faults as safeguards against systematic failure (man-made failures in specification, design, manufacturing, installation, operation, maintenance, and modification) and on adequate measures to control random failures (hardware faults due to the finite reliability of the components). Given that the CENELEC considers it unfeasible to quantify systematic failures, safety integrity levels are used to group methods, tools and techniques which, when used effectively, are considered to provide an appropriate level of confidence in the achievement of a stated integrity level by a system. The required safety levels connected with the 'man-made' unreliability are achieved through the satisfaction of standards of quality and safety management (CENELEC, 1999). The safety balance of the system is assessed through the concept of 'safety integrity levels', a measure of four discrete levels that enables the comparison of the qualitative and quantitative estimation of risks. The CENELEC standard provides tables where safety integrity levels correspond to intervals bands for hazard rates, which are the result of the estimation of the quantitative assessment. Safety levels and risk tolerability criteria depends on legislative principles, such as the Minimum Endogenous Mortality (MEM) or the French Globalement Au Moins Aussi Bon (GAMAB).

2.2.2 Human Reliability Assessment

Human reliability assessment is a discipline that provides methods for analysing and estimating risks caused by unavoidable human errors, as well as assessing how to reduce the impact of such errors on a given system. Three functions of HRA are identified as follows (Kirwan, 1994):
1. Human error identification: What errors can occur?
2. Human error quantification: How probable is it that the errors will occur?
3. Human error reduction: How can the probability that errors occur be reduced?

HRA is regarded as a hybrid discipline, founded on both a technical, engineering perspective (to provide understanding of the technical aspects of systems) and a psychological perspective (to provide understanding of the psychological basis of human error). The combination of these perspectives provides a foundation for assessing a total risk picture of a system, and to determine which factors impose most risk (human or technical).

HRA dates back to the early 1970s, when the nuclear industry developed systematic tools for the analysis and estimation of the operators' contribution to plant risk and safety. Today, there are many HRA methods available and several general approaches to HRA in the nuclear sector, with some being developed or adapted to other industries as well, such as petrochemical industry, aviation and air traffic management.

HRA has been used as a purely quantitative method, and human error probability (HEP) was defined as:

HEP = Number of times an error has occurred/Number of opportunities for an error to occur.

As we will see later, this quantitative approach was chosen to make HRA applicable to (quantitative) Probabilistic Safety or Risk Assessment (PSA/PRA). The quantitative approach was therefore necessary to ensure that human errors were included in the total risk picture. The focus has, however, shifted from a purely quantitative approach, as a consequence of recognising the importance of understanding the complexity and diversity of human error and its causes.

Independently of the general approach and industry domain, all human reliability analyses today are concerned with the variability of operator's action and in particular of those actions (or lack of actions) that may initiate or influence a system event in a positive or negative way. Unpredicted human performance variability, in fact, often becomes part of the causal generation of incidents and accidents (Hollnagel, 1998). In HRA, human-machine systems are analysed in terms of the interactions between hardware elements and human operators. In the case of hardware, errors are described in terms of safety systems failing to accomplish the functions they were designed to perform. As humans are involved, errors are represented as failures to perform particular functions at particular times. Many HRA methods further decompose these functions into basic tasks (such as reading an analog meter or diagnosing an abnormal event within 10 minutes, as in the HRA method THERP (Technique of Human Error Rate Prediction) that are associated with nominal failure probabilities, i.e. estimated failure probabilities before some environmental and personal factors (i.e. performance shaping factors) have been taken into account.

Although human tasks are specified in relation to basic psychological functions (e.g. observing, diagnosing) and contextual elements are considered in adjusting the failure probabilities (both environmental and personal), *in such descriptions the essence of human error remains the random human error variability associated with the basic task-function considered* – which can ultimately either be performed at the wrong time (error of commission) or not performed at all (error of omission). This is to say, human error, at its basic level of analysis, is random error.

One of the major criticisms of first-generation human reliability techniques can be stated in terms of the failure to give a proper answer to the causes of human error: the failure of explaining the causes of the human variability in performing the identified tasks would undermine the validity of the proposed human error probabilities. The criticism maintains, for instance, that in order to calculate the failure probability of diagnosing an abnormal event within 10 minutes, the nature of the diagnosis, its associated task complexity and attention demands are relevant, as well as the training of the operators, the availability of procedures, the task familiarity, and so forth.

However, the real issue issue should not be that the analysis stops at some basic task and its associated failure probability, but that without an adequate description of the psychological and contextual factors it is impossible to estimate meaningful failure probabilities for the tasks typically included in reliability analyses. In other words, if we had reliable human failure probabilities for basic tasks, and if the fault tree model was a valid model of a situation, then we should not bother with investigating the causes of, for example, failed diagnosis in terms of internal psychological error mechanisms, since the purpose of the analysis was a quantification of system risk calculated upon the consequences of combinations of hardware and human failures. The question remains whether (1) current human reliability techniques provide an adequate description of the psychological and contextual factors to enable estimation of meaningful failure probabilities and whether (2) event and fault trees are a valid way of modelling dynamic systems.

2.2.3 Classification

A market standard in the classification of human error in human reliability analysis is the scheme proposed in the early 1960s by Alan Swain (1963) (Table 1.4). Operator failures can be described in terms of:
1. Error of omissions: the operator fails to perform the required operation
2. Error of commission: the operator wrongly performs the required action
3. Extraneous errors: the operator performs an extraneous action.

The scheme refers to the behavioural level of human error in the performance of prescribed work tasks in a well-specified human-machine system. Operators are system elements that perform certain functions for the correct functioning of the overall system. The system design specifications provide the standard of correct functioning of humans as well as technical elements in all possible operating states. Without this premise of full predictably and possibly of predefined procedures for operators' tasks, it is not possible to distinguish between the three basic categories. For instance, it is logically impossible to distinguish between omission and commission since all commissions are by definition omissions of something (Hollnagel, 1998). Analysts classify malfunctions in the context of the prescribed functions in a

Table 1.4 Categories of incorrect human outputs (Swain & Guttmann, 1983).

Category	Sub-category	Examples
Errors of commission	Omits entire step	
	Omits a step in the task	
Errors of commission	Selection error	Selects wrong control Mispositions control (includes reversal errors, improperly made connections, etc.) Issues wrong command or information
	Error of sequence Time error	Too early Too late
	Qualitative error	Too much Too little

practical fashion. In particular, errors of commission are distinguished from errors of omission based on the assumption that a required task is initiated but incorrectly carried out.

2.3 Information processing

The roots of the information processing paradigm are the animal learning tradition, the stimulus-response behaviourist paradigm championed by Watson, communication theories, and the cognitive programme derived from Tolman's stimulus-organism-response model. The central view in the paradigm is that humans are processors of information and psychology's task is to discover and describe the functions, mechanisms, stages, and limits of the human processing capabilities. The information processing metaphor is clearly analogous to the technical model used since the 1940s for the description of the first computers and later extended to all automated systems. Concepts and vocabulary are permuted from this tradition and from communication theory. Examples are central processor, working memory, bottleneck, overload, and capacity limits.

Information processing represents human cognition as a process where a flow of information begins with a stimulus, passes several stages of processing (e.g. perception, attention, memory storage), which transforms the information demanding some time, and results in a response. This representation maintains the stimulus-response assumptions of (a) a defined sequence and (b) a basic logical-temporal dependence on a stimulus. At the same time, it elaborates the 'o' of the stimulus-organism-response paradigm by formulating empirical hypotheses regarding the proprieties of the processing systems and the communication channels.

The information-processing paradigm has developed in two directions, a quantitative and a qualitative one. The quantitative tradition has been conducted in the laboratory within the natural science tradition of experimental psychology. It has focused on the resource-limited aspects of human cognition: attention and memory. Theories are advanced to explain well-defined and manipulatable phenomena, such as learning lists of nonsense syllables to quantify short-term memory, or presenting different messages to different ears and asking the listener either to repeat all the words or to monitor the message of one channel in order to investigate parallel processing and interference. The drawback of these types of studies is that there is always a very strong trade-off between internal and external validity, that is, while contrasting theories can well explain controllable and paradigmatic situations, none of them tell us much about of how people work in more complex, real tasks. Hollnagel defines this line of research as micro-cognition, where the 'emphasis is more on predictability within a narrow paradigm than on regularity across conditions, and more on models and theories that go in depth than in breadth' (Hollnagel, 1998). *Human errors in this tradition are the result of processing limitations, and occur when task demands exceed capacity.*

The qualitative tradition has emphasised the process of cognition rather than the resources limitations. Models have elaborated on the stages of information processing, such as diagnosing and decision-making, in more complex situations than those analysed in the quantitative orientation. Qualitative information processing has been carried out within the cognitive tradition. Although this is a very recent discipline, without generally established aims and principles, it is probably correct to locate as central themes (1) the representation and application of stored knowledge, (2) the study of realistic tasks where the interaction with the environment is taken into account, and (3) the consideration of controlled, conscious processing modes as opposed to automatic, unconscious ones.

The cognitive tradition became popular in the mid-1970s when a number of different research trends in psychology, namely the work of Minsky (1975) on computer vision, Rumelhart (1975) on the interpretation of stories and Schmidt (1975) on motor skill learning, reintroduced Bartlett's concept of schema (Bartlett, 1932) and the active role of cognition initiated by the Gestalt tradition. These theories subscribe to the principle that the total is more than the sum of is parts, thus rejecting an atomistic view of mental processes. Despite differences in language, schema theorists agree that there are high-level knowledge structures already active at the early stages of information processing. Perception, for instance, is not a passive collection of external bits of information but is dependent on the preconceived knowledge structures that anticipate much of what will occur, and which are activated by combinations of external and internal triggers. Missing features of the environment can thus be provided by memory or can simply assume default values. Errors can occur either by the activation of wrong schemata or by the faulty assignment of values to some schema variables. As this process is not limited to perception, but concerns all cognitive activities, *human errors can generally be conceived as over-adapted responses, a psychological tendency to overrely on past experience.* The idea of 'default assignment' is the central thesis of Reason's influential model of human error.

The most referred to information processing model is that of Wickens (1992) (Figure 1.2). Wickens' model is a good illustration of the orderly sequence of stages in the information processing paradigm. The starting point is the stimuli from the environment that are received by the senses. The sensed stimuli are initially processed through sensory modality specific (e.g. visual, auditory, kinaesthetic) short-term sensory stores (STSS) where the representation of physical cues are prolonged for a short period after the stimulus has physically terminated. Wickens attributes the following properties to the sensory stores: (1) they do not require conscious attention to prolong the presence of the stimuli after these have ceased, (2) they are relatively veridical, meaning that they preserve details of the stimuli, and (3) decay times are dependent on sensory modality, but are always rapid (one to eight seconds).

Perception concerns integrating and assigning meaning to the sensory inputs. The most basic form of perception is detection, which concerns determining whether a sign or target is present. At a higher level of processing the targets are assigned to the class they belongs to – a process referred to as recognition or identification. Perceptual judgements are distinguished between absolute and relative. Absolute judgement concerns identifying a stimulus on the basis of its position along one or several stimulus dimensions (e.g. length of an object), whereas relative judgement concerns determining the relative difference between two or more stimuli (e.g. which object is longer).

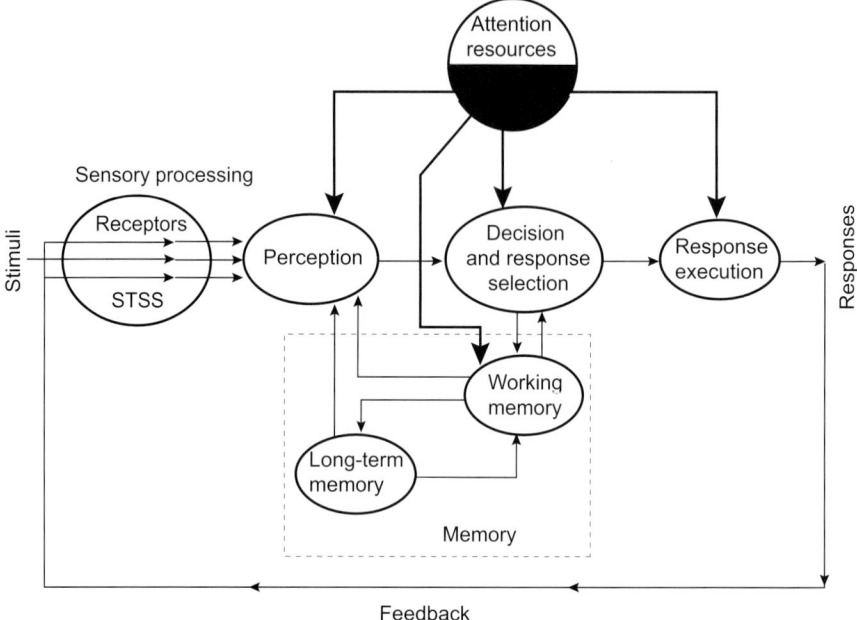

Figure 1.2 Wickens' model of information processing.

The next stage in the model is called decision and response selection and occurs after meanings have been assigned to the physical cues (a stimulus becomes information once it has been assigned meaning). Decisions may be rapid or thoughtful, and the individual may choose to execute a response. Alternatively, information may be stored in the memory for a short period (seconds to minutes) in working memory, by rehearsal. According to Baddeley and Hitch (Baddeley & Hitch, 1974), working memory consists of three stores:

1. A modality-free central executive – this has limited capacity and is used when dealing with cognitively demanding tasks.
2. An articulatory loop (a 'slave system') – this holds information in a speech-based form.
3. A visuospatial scratch pad or sketch pad (a 'slave system') – this is specialised for spatial and/or visual memory.

Information can be transferred for a longer period (hours to years) in long-term memory by learning.

The model distinguishes between the decision to initiate a chosen response and its execution. The latter phase is denoted response execution. In this phase errors are typically associated with problems of automaticity, which refers to the fact that people are able to execute highly practiced action sequences with little attention. Such activities are associated with a specific type of error, namely slips.

The outcome of the decision can function as a basis for further processing of cues. This is represented by the feedback loop in the model, which starts following the response execution. However, the information flow does not always start with external stimuli, but can be triggered internally as, for example, by thoughts in working memory. In addition, the flow does not always follow a left to right direction, as in the case where immediate experience represented in working memory affects perception.

An important aspect of information processing models is that the stages of perception, decision-making, and response selection and execution are, as illustrated in Wickens' model, largely dependent on the available attention resources. It is hypothesised that limited amounts of attentional resources exist that can be distributed among the different processing components.

2.3.1 Classification in information processing

As an example of classification in the information processing tradition, I present the Human Error Reduction in ATM (HERA) technique, developed by EUROCONTROL within the scope of the EATCHIP/EATMP (European Air Traffic Control Harmonisation and Integration Programme/European Air Traffic Management Programme) work programme.

2.3.1.1 Model

HERA is based on Wickens' (1992) model of information processing. However, a number of modifications were introduced to tailor it to air traffic management (ATM) and to resolve various criticisms. Working memory follows from perception and contains the controllers' mental representation of the traffic situation. A 'self-picture' is introduced, i.e. thoughts that controllers have about themselves and their ability to cope with the traffic situation. In addition, decision and response selection are two separate processes (Figure 1.3).

2.3.1.2 Classification scheme

The classification system of HERA considers two factors:
- The error, i.e. what error occurred (type), how did the error occur (mechanisms)?
- The context, i.e. when did the event occur, who was involved and what was their involvement (including the organisation factors), where did it occur, what tasks were being performed, how did the event occur, and what information or topic did the error involve?

HERA employs a quadripartite distinction of error categories: Error/Violation Types (ETs), Error Detail (ED), Error Mechanisms (EMs) and Information Processing Levels (IPs).

The Error/Violation Types (ETs) are the way the action manifested itself externally. In order to decide what is the 'right' or wrong' action all relevant procedures and expected actions will be considered. ETs include errors of omission, timing, sequence, quality, selection, and communication; examples are: omission, action

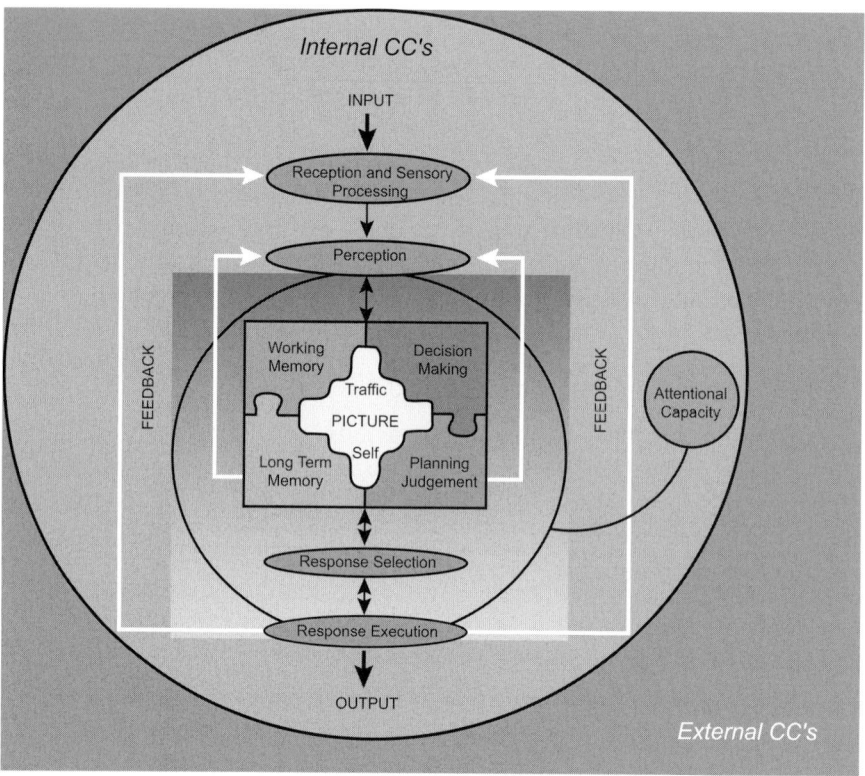

Figure 1.3 HERA model of information processing (Isaac et al., 2002)

too late, mis-ordering, extraneous act, right action on wrong object, and incorrect information transmitted.

Procedural violations are actions that contravene a rule, procedure or operating instruction. Procedural violations are more complex than errors and are defined in terms of procedures, controller intention and awareness, and working practices and circumstances.

Error Detail (ED), and the Error Mechanism and the Information Processing levels, all describe the error from a psychological perspective. There are four ED domains covering all the information processing activities:
- Perception and vigilance
- Memory – working and long-term
- Planning and decision-making
- Response execution.

The EDs classify the error at a gross level (e.g. error of 'working memory' or 'response execution') and direct the user to a subset of errors within the relevant ED domain – the Error Mechanisms (EMs). The Error Mechanisms (EMs) describe the internal manifestations of the ED (e.g. misidentification, late detection, misjudgement). They refer to the cognitive function that has failed. The Information Processing Levels (IPs) describe how the psychological cause influences the Error Mechanism EM within each Error Detail (ED) level. These 'psychological causes' refer to inherent human fallibilities which influence behaviour, such as visual discrimination, expectations, working memory capacity, confusion, and habit. For example, the IPs within the ED 'perception and vigilance' include 'expectation bias' (i.e. seeing or hearing what one expects to hear), 'information confusion' (i.e. confusing two things that look or sound alike), and 'distraction/preoccupation' (i.e. temporary interruption by an external event or more prolonged loss of concentration due to internal thoughts).

There are three categories in HERA that describe the context at the time of the error: the Task, The Information and Equipment, and the Contextual Conditions (CCs). The Task describes the function(s) that the controller was performing at the time the error was made. Example of such tasks include: coordination, tower observation, planning, RT communications and instruction, control room communications, strip work, materials checking, radar monitoring, HMI (Human Machine Interface) input and functions, handover/relief briefing, takeover, training, supervision, and examination. The Information/Equipment lists describe the environment in which the error occurred. Examples of HERA information/equipment elements are: procedures, coordination, aircraft type, geographical position, airport, flight rules, secondary radar, visual approach aids, aerodrome equipment, flight information displays, and input devices.

Contextual Conditions (CCs) can be defined as factors, internal or external to the controller, which influence the controller's performance of ATM tasks. Contextual Conditions (CCs) can help to explain why the error occurred. CCs include the following sub-categories: pilot-controller communications (e.g. pilot breach of RT standards/phraseology), pilot actions (e.g. responding to traffic alert and collision avoidance system (TCAS) alerts), traffic and airspace (e.g. excessive traffic load/complex traffic mix), weather (e.g. extreme wind at high altitude), documentation and procedures (e.g. inappropriate regulations and standards), training and experience (e.g. controller under training), workplace design, HMI and equipment factors (e.g. RT failure), environment (e.g. lighting – illumination, glare), personal factors (e.g. high anxiety/panic), team factors (e.g. poor/unclear coordination), organisational factors (e.g. problems in the work environment), and administrative workload problems. There may be more than one CC for an error. Contextual factors are also important for the creation of an error database.

2.3.1.3 Method

The analysis uses incident investigation reports, proceeds from the beginning of the description and moves forward in time, and creates a description of how human errors propagate and result in incident. The classification system of HERA was developed in two forms, a tabular format and a series of decision flowchart diagrams. Tables are available for Tasks, Information and Equipment, Error/Violation Types (ETs), and Contextual Conditions (CCs). The diagrams allow analysts to identify errors by answering a series of 'Yes'/'No' questions. There are separate decision flow diagrams for: Error Detail (ED), Error Mechanisms (EMs) for each error detail, Information Processing levels (IPs) for each error detail, and Contextual Conditions

Embodied Minds – Technical Environments

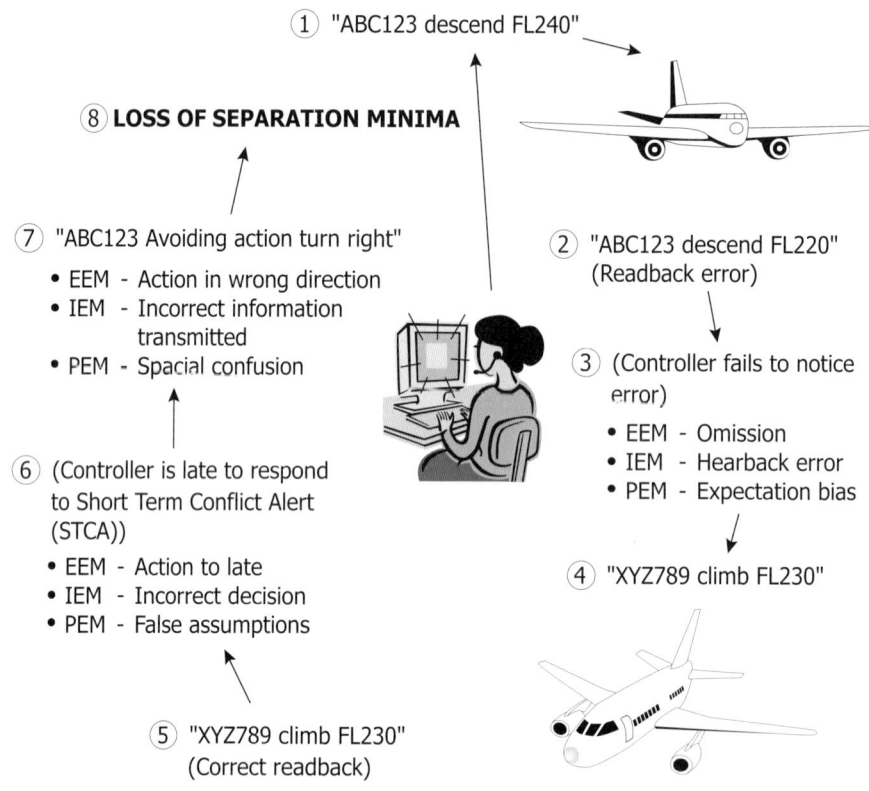

Figure 1.4 Pictorial description of an ATM incident and associated HERA error classifications (Isaac et al., 2001b).

(CCs) sub-categories. Each decision flow diagram starts at a different Error Detail (ED) domain. This allows the analyst to start at the applicable ED and makes the technique more resource efficient. The decision flow diagrams allow the analyst to begin at any ED domain. Also, the format allows the analyst to skip ED domains where they are confident that the error did not occur within that area, or where the analyst is directed to 'jump' to another ED domain by following the diagram.

HERA's internal structure of Error/Violation Types ETs, Error Mechanisms EMs and Information Processing Levels IPs allows the analyst or incident investigator to classify errors at three levels of detail. There should almost always be sufficient information to classify the ET, and usually there will be enough information to classify the EMs. IPs add value to the analysis, but are the most difficult 'level' to classify because there is sometimes insufficient information to determine the psychological cause (see Figure 1.4 for a pictorial description of an incident).

2.4 Cognitive System Engineering

Cognitive System Engineering (CSE) implies a distinctive approach to human performance and human error in complex human-machine systems. Cognitive system engineering is a theoretical perspective on human-machine systems design and human performance modelling that has been advanced in the works of Erik Hollnagel and David Woods (Hollnagel, 1993; Hollnagel & Woods, 1999; Woods, 1986; 1988). Cognitive system engineering is a response to the rapid technological progress produced by the digital revolution. Since the late 1970s, the evolution of machines towards becoming 'intelligent' agents as well as the occurrence of certain events, such as the Three Mile Island incident, demonstrated the deficient status of man-machine systems knowledge at the time.

In cognitive system engineering the human factor problem of assisting performance is translated into the one of achieving human-machine integration, where the role of machines has changed due to increased computer power and the potentialities of artificial intelligence.

The central tenet of this perspective is that contemporary human-machine systems are best viewed in terms of joint cognitive systems, and should therefore be designed, analysed and evaluated as such. A physical or biological system is considered a cognitive system when it satisfies the following criteria (Hollnagel & Woods, 1999):

1. Its behaviour is goal oriented
2. Its behaviour is based on symbol manipulation
3. It uses knowledge of the world
4. It is adaptive to new circumstances and can see problems in more than one way.

A cognitive system possesses knowledge about itself and the environment and is able to plan and modify its actions on the basis of that knowledge. In this sense it is concept driven, in contrast to data driven non-cognitive systems whose actions are simpler responses to stimuli. Concept-driven behaviour is produced by means of models of representations of the environment. Humans are clearly cognitive systems and use models of the environment in intelligent behaviours such as planning and deciding. Human-Machines systems are also cognitive systems, even when the machine part alone does not satisfy all the criteria to qualify.

The operators and the process or machine have to be modelled on equal terms and the interaction between the human and machine parts of the human-machine cognitive ensemble is the focus of CSE. In particular, this interaction is thought to be more complex and dynamic in the way the environment is set to interact with the human operator in the information processing model (i.e. signals, executions, disturbances). In the words of David Woods (Woods, 1988, p. 153):

> The configuration or organization of the human and machine components is a critical determinant of the performance of the system as a whole. The joint cognitive system paradigm demands a problem-driven, rather than a technology-driven, approach where the requirements and bottlenecks in cognitive task performance drive the development of tools to support the human problem solver.

There are two extremes in the way this interaction can be configured. Intelligent machines can be used as tools that expand human capabilities. The machine amplifies the fundamentally correct human capabilities, overcoming some of their limitations (e.g. memory, attention resources). Human performance is shifted to a different yet higher level. In a man-machine system designed in this way the locus of control is the human part. At the other extreme, machines are conceived as prostheses, replacements or remedies for human limitations. The machine compensates for the deficiencies in human problem solving and reasoning. In a cognitive support system designed in this way the locus of control of the joint system is the machine, while the human is the data gatherer and action implementer of the stand-alone problem-solver machine.

While the latter view of machines as prostheses is sometimes endorsed by information processing approaches, cognitive system engineering is definitely oriented towards the former view of machines as tools. The joint cognitive system perspective defines a computer consultant as follows:

> [A] reference or source of information for the problem solver. The problem solver is in charge; the consultant functions more as a staff member. As a result, the joint cognitive system viewpoint stresses the need to use computational technology to aid the user in the process of solving his problem. (Woods, 1986, p. 161)

The consultant does not provide a solution plus a solution justification, but performs the role of a good advisor: 'Good advisory interaction aid problem formulation, plan generation, (especially with regard to obstacles, side effects, interactions and trade-offs), help determine the right question to ask and how to look for or to evaluate possible answers' (Woods, 1986, p. 160).

Methodologically, CSE investigates human performance on a global level and in realistic tasks. The focus is on the overt phenomena, on how tasks (controlling a process, conducting a car) and cognitive activities (such as planning and decision making) are performed and achieve their goals, rather than on the underlying psychological mechanisms of the cognitive activities implied. CSE can thus be defined as ecological, in the sense that the study of human performance and problem solving behaviour is considered meaningful only in relation to the tools (i.e. support systems) and the world that drive the behaviour under study. This is in contrast with most laboratory research which analyses problem solving without tools, and hence with the risk of eliminating critical features of performance (Woods, 1988).

2.4.1 Human error in CSE

The problem-driven approach and the view of support systems as tools lead to the second central assumption of the cognitive system engineering perspective: human cognition is an active process dependent on two equally strong determinants, the operator's goals and the situation or context. The influence of the environmental circumstances, the context of performance, is the core of the description and analysis of human error, as performance failures are not traced back to the internal information processing malfunctions but interpreted as mismatch between cognition and working conditions. This mismatch is inevitable in complex systems and it is due to the unanticipated variability that characterises all highly dynamic and highly coupled worlds. Unanticipated variability is the result of 'underspecified instructions, special conditions or contexts (violations of boundary conditions ... or impasses where the plan's assumptions about the world are not true), human execution errors, bugs in the plan, multiple failures, and novel situations (incidents not planned for)' (Woods, 1988).

Human error in CSE, or in Hollnagel's terms, erroneous human action, is thus the consequence of two determinants: the human-machine mismatch and the inherent human variability. Human-machine mismatch can thus, in principle, be eliminated or minimised through design, although it is impossible to predict all possible operating situations in advance and therefore no totally matched man-machine system is possible. On the other hand, even a hypothetically completely matched system will not be free from human errors since there always remains the possibility of residual erroneous actions, which are due to inherent human variability, i.e. 'random fluctuations in how the mind works, subtle (or even less subtle) influences from the environment, forgetting, loss of attention, associative jumps, etc.' (Hollnagel, 1993). Since residual erroneous actions and knowledge mismatches naturally blend in daily operations it is empirically difficult to isolate them. This would create a problem for a quantification of human error if performance were decomposed into assumedly basic actions outcomes.

It follows that the ideas of failure mechanisms and processing limitations that are central in the information processing approach to human error are considered unnecessary. In the first place, these mechanisms are hypothetical constructs. In the second place, they can be studied only in isolation or laboratory tasks (micro-cognition). Finally, the contextual effects on human performance are underrepresented and underestimated, as a result of the previous two points. An alternative approach for the classification of the causes of human erroneous actions and the assessment of human reliability is developed by adopting a more holistic approach that starts with a thorough evaluation of the global context of performance. In this case, the situation and the task are set to determine the goals of the agent, which in turn governs the control of actions as well as the number and sequence of cognitive functions that will be active in the process. Each resultant control mode active in the particular task is then associated with an estimated level of human reliability or intrinsic performance variability.

For CSE, the response to unanticipated variability is not creative problem solving but coordination by resource management. This is a process where background knowledge is gradually unfolded in the process of monitoring and adapting plan execution in reaction to deviations from pre-planned responses. CSE individuates a series of strategies to help the users deploy the background or meta-knowledge they possess, or in other words, to enhance demand-resources match:

(1) The study of human-human cooperation, where the machine is thought of as a cognitive agent.
(2) The supervisory control model, where the human has ultimate authority. Shared representation is required by the two agents, as well as the supervisor must be able to redirect the lower machine.
(3) The view of machines as extensions and expansions, where people are tool builders and tool users. Tools magnify the capacity for physical work, perceptual

range and, as CSE is particularly concerned, cognitive environment. The latter is enhanced by calculation power, search and deductive power, and economy of representation in terms of conceptualisation power. This is the ability to experiment, to visualise the abstract, and the enhancement of error tolerance by feedback about effects and/or results.

2.5 Risk management models

By risk management models I indicate a class of approaches that in the literature go under the various names of error management (Bove, 2002), threat management (Helmreich et al., 1999), resource management (Amalberti, 1992), and control of danger (Hale & Glendon, 1987), to name a few. A common denominator in these approaches is that they pursue system safety without concentrating on errors per se but on the generation and propagation of system hazards, including human-generated hazards, and on the way these are prevented from resulting in accidents. In describing such processes, risk management approaches take into account the whole process of successful and unsuccessful performance, and clarify how hazards and errors are anticipated and detected, how they are recovered, and in general, how humans achieve and maintain control in risk situations.

Similarly to human cognitive engineering these approaches underlie the active and anticipative role of the individuals in their strategies to maintain safety. Amalberti (1992) has proposed a risk management model, or 'model of anticipation', to describe the way individuals control safety-critical processes. At the core of the model are anticipation and action, as the means by which risk, which is a normal task component and part of operators' competence, is managed. Operators know their resources are limited, and are therefore forced to take risks. However, they have a representation of their competence (meta-knowledge) that allows them to continuously maintain risk at the lower level compatible with expected performance. They thus develop strategies to adapt to task demands, i.e. to manage risks, to maintain performance within safety margins, to switch between short-term activities and long-term reasoning, etc. These strategies typically involve two stages. The first stage involves anticipating the course of the situation. Based on simplified heuristics (e.g. application of known schema, hypothesis testing, saliency, and frequency of occurrence) and previous experience, actions are continually taken from a repertoire of known solutions in order to force risky situations within known safety margins, i.e. within pre-planned models of the situations. It is clear at this stage that risks are managed by taking other risks. The second stage is directed at the definitive solution of the problem with the identification of the causes of the deviations. Depending on the success of the first stage, the risks are finally minimised at the second stage. The two stages involve different cognitive mechanisms: the first stage involves primarily skill and rule-based reasoning but reactive behaviours are kept to a minimum in applying prepared responses and escaping, 'freeing time in order to optimise response' (Amalberti, 1992, p. 103). The second requires pre-eminently knowledge-based reasoning, and requires time for elaboration and mental effort.

The error classifications inspired by this class of models are not conceptually different from those based on information processing. That is, the human error is still a cognitive category, while the difference rests on the fact that the classification is not focused on the human error per se, but rather on the entire human performance in safety critical situations. In other words, these approaches (similar to CSE) are coupled tightly to a system safety model that assigns a role to the human performance that is as much interested in explicitly modelling the positive contributions to safety as it is the negative ones. A very well developed classification system based on an error recovery model is the one developed by Bove (2002).

2.6 Violation models

We have seen that the conceptual difference between errors and violations is less clear-cut than one might intuitively assume. The fact that a liberal conceptualisation of the concept of violations allows to the inclusion of a large number of unsafe acts has been exploited to develop alternative approaches to the study and management of systems safety. These perspectives parallel the traditional study of human error insofar as they also identify few external manifestations of violations and a more articulated set of their causes. There are important differences, however.

First, the standards of correct performance, which define the manifestations, are the safety rules of given organisations. Hence the relation to unwanted consequences, such as accidents, is only indirect (as non-compliance does not necessarily lead to system failures). Second, the causes of violations are not only identified based on cognitive theories of human performance but mainly also on social psychology frameworks (see Table and Section 1.7).

The most important consequence of this change of perspective is that classification systems for violations are (in their present state) not so much tools for learning from the past and predicting future human contributions to system failures, but more theoretical models to support techniques and methods that can help mapping and reducing the potential for a special class of human hazardous behaviour: individual and group violations.

2.6.1 Causes of violations

The causes of violating behaviour point to the motivational, attitudinal and social dimensions of human behaviour. In contrast to cognitive approaches, violation-based approaches for the reduction of system accidents and damages do not focus on the individual's elaboration of information. Instead, they concentrate on the motives, attitudes and intentions relating to violating existing rules and safe practices. Rather than looking for psychological 'error mechanisms', they investigate the group, organisational and social determinants of decisions and behaviour. The accent is then on why individuals do not follow rules and procedures, on why prescribed tasks and redefined tasks diverge, and on the factors that influence such redefinitions.

Different theories of safety rules non-compliance have been put forward (Collier 2000, Mason 1997, Hale et al. 2003, Lawton 1998; Reason et al. 1998). They point out a host of different reasons for people not complying with rules and regulations, ranging from individual motives such as saving time, and technical causes such as poor ergonomic design, to organisational issues such as lack of inspection and training. In the following sections I will describe the MTO framework of non-adherence behaviour and a classification system proposed by Mason (1997). First, however, the issue of violations and goal conflicts is described.

2.6.2 Violations and goal conflicts

As seen in Section 1.7, violations can be 'reduced' to taxonomies of error, provided that the models behind such taxonomies are able to treat a wide spectrum of cognitive activities. In particular, in order to classify violations as cognitive errors, one needs the theoretical tools for dealing with high-level decisions, as the basic question is why available rules are not judged adequate for particular situations.

A psychological treatment of the issue of violations may rely on the principles of goal conflict. Given that human behaviour is goal-directed and governed by knowledge activated in situations, clues are available from looking at people's goals at the time in question. It is seldom the case that just one goal governs what people do. Most complex work is characterised by multiple goals, all of which are active or must be pursued at the same time. Depending on the circumstances, some of these goals may be at odds with existing safety rules, or different existing rules might imply conflicting goals.

Lewin (1935) defines conflict as 'a situation in which oppositely direct, simultaneously acting forces, of approximately equal strength, work upon the individual'. In theory, the opposition between two forces in a conflict situation leads to an increase in tension, which has negative implications for performance (Kahn et al., 1964). Hellriegel et al. (1998) state that: 'Goal conflict refers to incompatible preferred or expected outcomes. Goal conflict includes inconsistencies between the individual's or group's values and norms (such as standards of behaviour) and demands of goals assigned by higher levels in the organisation'.

Two classes of real or perceived goal conflicts can be identified:
1. Between the individual and the organisation
2. Within the organisations' goals system.

An example of conflict between the goals of the individual and the organisation is when individual and organisation are maximising different and conflicting things (Hollywell & Corrie, 2000). For example, while an organisation may wish to constantly maximise safety, an individual might decide, in particular situations, to maximise productivity, speed, comfort, financial benefit, or social conformance.

However, there are cases where individuals commit violations in order to maximise safety. In extreme and abnormal cases, rules and procedures might not be adequate, applicable or even available. Examples of this kind of situation can be found in the nuclear domain (e.g. the Davis-Besse incident; see Kirwan, 1994) and from transportation (e.g. the Clapham Junction incident; Hidden, 1989). Hollywell

& Corrie (2000, p. 3) maintain that '[I]t can be argued that in particular (and hopefully, extremely rare) situations the violation of a rule or procedure could lead to increased safety'. They point to the fact that in those situations operators have to choose between obeying a rule or using their knowledge of the system, and reflecting on the effect of recent changes in the UK Railway system, they conclude that one 'cannot have both prescription and adaptability in the same context; one approach has to be chosen to manage safety' (p. 3). This conflict might not be as clear-cut as it first seems, as all safety relevant systems contain both elements, yet the problem of the optimal configuration of prescription and adaptability might be considered rightfully one of the major challenges in safety-systems design.

One supporter of the adaptability alternative is Ruiz Quintanilla (Ruiz Quintanilla, 1987), who claims that allowing operators the possibility of individual choice of problem solving strategies would result in the avoidance of 'monotony, satiation and error proneness' (p. 127), and, in this context of violation one could add, to the reduction of violating behaviour. Ruiz Quintanilla grounds his claim in psychosocial research on the meaning of working in different social groups. Workers identify with their work and profession provided they dispose of discretionary freedom and the potential for self-regulative activities within the organisation. System and job design should take into account not only the cognitive capabilities of the operators but also their preference structures and job expectations, and in general, all their work-related value orientations.

Connected with the dilemma between prescription and adaptability is the role of automation in socio-technical systems. Automation can constrain the individual's safety control strategies and plans by introducing a system rigidity that contradicts human flexibility and anticipation. In this respect, violations take the form of overriding the automaton by deactivating some protection systems in order ensure safety on the individual's own premises. This kind of violation is more generalised than one would expect, as Amalberti (1992, p. 104) found in the aviation context:

> [S]trategies of detouring systems from their standard uses are quasi-systematic with expert pilots, but with various levels of frequency and risk taking (in a questionnaire, 86% of a population of fighter pilots responded that they were using such strategies frequently or fairly frequently).

A third related potential source of conflict between the individual and the organisation lies in the justification of rules and regulations: although their formulation attempts to ensure safety and other systems' goals, their efficiency is not necessary self-evident. That is to say, as there is no guarantee that following procedures in any case will guarantee the system goals, there is likewise no guarantee that not following them will result in negative systems consequences. The links between regulations and systems negative outcomes are at best logical and empirical and at worse taboos and superstitions. Insofar as individuals will retain the freedom to choose behaviour, the potential for rule violations will always be present, even to the point of becoming normal behaviour. It is well documented in the literature that informal group norms and behaviours develop as result of working groups interactions with technical systems, and these might deviate from the norms prescribed by the designer or the organisation.

The shift of attention from the individual to the social dimension of unsafe acts makes a violation approach particularly attractive from the point of view of the measures an organisation can put in place to reduce the potential for undesired consequences. It is clear, in fact, that violations are strongly influenced by company management; they are created by, and accepted or condoned by management.

2.6.3 MTO framework of non-adherence behaviour

The Man-Technology-Organisation framework can be utilised to classify the causes of non-compliance behaviour under the three elements of the system. We will then have individual level causes (Man), working conditions (Technology) and organisational causes, as exemplified in Figure 1.5.

The MTO framework is both a theoretical view of complex systems as well as a practical tool for safety management. The latter aspect is especially relevant in the case of safety rules compliance. To illustrate the point, we can refer to The IFE Safety Rules and Compliance Questionnaire (Massaiu & Kaarstad, 2007), where

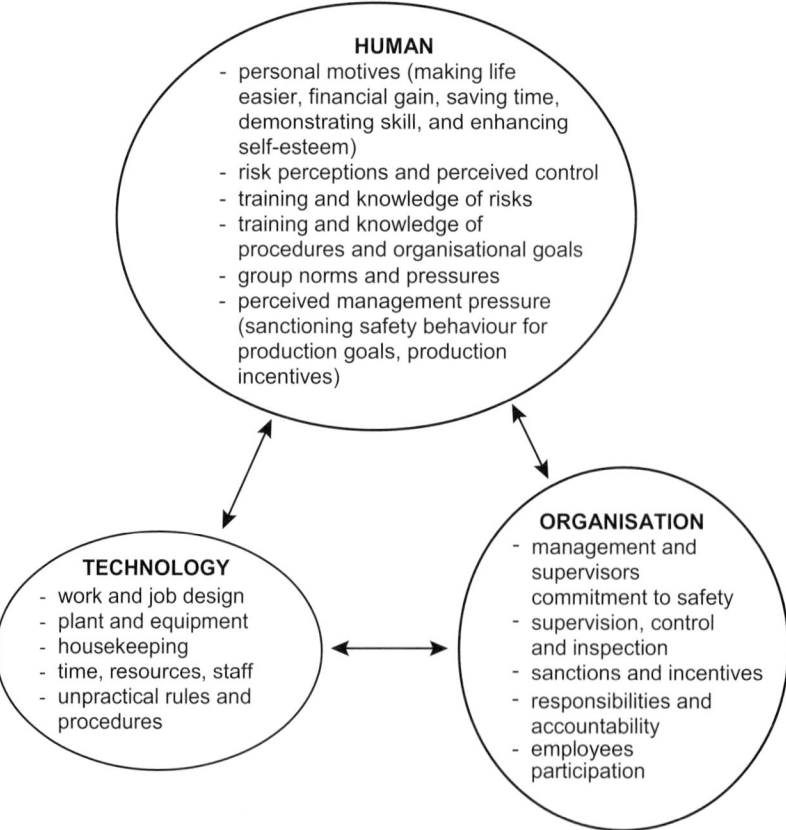

Figure 1.5 Causes of non-compliance in the MTO framework.

the three classes of assumed causes are broken down into the following dimensions:
1. Motives
2. Risk Perceptions
3. Group Norms
4. Management and Supervisory Attitudes
5. Safety Status
6. Technology and Working Conditions

Motives for not following safety rules will typically be time and financial gain, less effort, gaining prestige and social acceptance, as well as enhancing self-esteem and demonstrating skills and expertise (powerfulness).

Risk perceptions are connected to the individual's feelings of control of hazards. Both rational and emotional dimensions of risk perceptions should be considered as well as complacency caused by accident-free environments, good safety records and repeated positive feedbacks of unsafe behaviours.

Group norms refer to safety attitudes and practices of the workforce and can be operationalised as pressure from authority figures and peers to break the rules, and lack of commitment to safety of co-workers and trade union representatives.

Management and supervisory attitudes includes pressure to prioritise production, and lack of commitment to safety in terms of lack of safety programmes, condoning unsafe behaviour, lack of support for reporting mistakes and concerns, lack of response to workforce concerns (e.g. unpractical procedures), lack of role models such as leading by example, and lack of feedback to workforce on effectiveness of safety measures.

Safety status refers to organisational states and conditions such as inadequate supervision, ineffective disciplinary procedures, poor accountability, lack of competence and training, lack of participation and cohesion, and dissatisfaction with other people.

Technology and working conditions might pressure rule violations. Typical examples are unpractical rules and procedures (emergency, operating instructions, maintenance schedules)and their availability, difficulty in language, format, and logic; conflicts between procedures, unavailability of tools and equipment, poor ergonomic design of interfaces, controls, plants; poor housekeeping, poor resource planning, and excessive workloads.

This modelling uses very broad categories, under which comprehensive lists of possible causes of violating behaviour suggest feasible indicators. The result is a tool for assessing the degree of compliance and non-compliance to the safety rules of an organisation, as well as for investigating the reasons for such behaviour.

2.6.4 Violation and risk taking behaviour

Performing tasks in a manner that violate rules and regulations, which are introduced to guarantee the safety of the socio-technical systems, would be expected to affect the level of risk connected with such activities. It is therefore not surprising to find individuals' judgments and perceptions on risk and safety as determinants of violating behaviour. However, the practical usefulness the concept of risk taking in this context is problematic.

In road traffic research the concept of risk compensation has been introduced to explain why certain safety measures achieve less than expected results. According to the risk compensation theories, an individual perceives a level of risk connected with a particular activity and adapts his/her behaviour in order to reach certain personal goals with a certain target level of safety (Wilde, 1974). For example, the motivation for saving time may imply that road improvements (e.g. wider lanes or shoulders) result in higher speeds, with no net effect on safety as a consequence.

However, the idea of perceiving and weighing risks in safety-relevant behaviour is controversial. For instance, Taylor (1987) claims that incidents occur as a result of risk taking behaviours, but these are not rational decision-making situations. Taking risk, in this perspective, is not balancing situations, as negative consequences cannot be weighted. The stochastic property of accidents and 'divergence' in catastrophe theory, i.e. many consequences from the same antecedents cause negative stochastic events to be perceived as 'black holes', something incomparable and without meaning, in contrast to positive stochastic ones, which motivate the intentions to take risks. Instead, negative consequences take the form of the following:

> [F]ears of non-conformity, to respected opinions, social deviance, or lawbreaking. A driver venturing on a short journey at night without functioning lights probably does not feel that his life is at risk, but may well fear that the police will stop him, or that other drivers will protest, or that if an accident does happen he will be made to take the blame, or even that to behave in such a way is to act in a foolish and uncaring manner inconsistent with his image of himself. The balance may, in other words, be a moral one. (Taylor, 1987)

This 'balance' might alternatively be considered a motivational and attitudinal one rather than a moral one, and theoretical explanations could be provided in psychological terms, although not prominently cognitive ones. At any rate, it should be stressed that although risk perceptions and attitudes are probably present in all violating behaviours, there seems to be a much richer combination of determinants involved in the choice and execution of 'redefined' tasks, at least insofar as many such redefined tasks would not be perceived as changing (and even would not change) the level of risk connected to the activity.

2.6.5 Classification in the violations framework

Mason (1997, p. 288) describes violations as 'deliberate deviations from the rules, procedures, instructions or regulations introduced for safe or efficient operation and maintenance of equipment', and estimates that 'up to 90% of accidents occur when an individual, or individuals, deliberately contravenes established and known safety rules' (Mason, 2000).

He has proposed a classification of violation based on the factors that influence a person's decision to break rules. These are considered at two levels: factors that motivate the violation and factors that influence the decision to violate. Factors of the first type are called 'direct motivators', which directly motivate management and operating and maintenance personnel to break the rules. Factors of the second type are named 'behaviour modifiers', which may increase or reduce the probability of any individual deciding to commit a violation. Table 1.5 lists the direct motivators and behaviour modifiers identified by Mason.

Mason's classification includes both individual and organisational factors in both classes of factors, but the emphasis is on the organisational level.

The UK's Human Factors in Reliability Group (HFRG) and the UK Health and Safety Executive (HSE, 1995) jointly published a methodology for addressing procedural violations that is grounded on Mason's classifications of direct motivators and behaviour modifiers. This approach identifies organisational factors that increase the potential for violations. Such organisational factors include safety commitment, training, management and supervision, job design, and equipment design. The approach can be applied by the non-specialist and is applicable to a wide range of industries.

Table 1.5 Mason's classification of violations.

Direct Motivators	Behaviour Modifiers

• Making life easier
• Financial gain
• Saving time
• Impractical safety procedures
• Unrealistic operating instructions or maintenance schedules
• Demonstrating skill and enhancing self-esteem
 	
There could also be:
• Real and perceived pressure from the 'boss' to cut corners;
• Real and perceived pressure from the workforce:
  (a) To break rules,
  (b) To work safely.
	• Poor perception of the safety risks
	• Enhanced perception of the benefits
	• Low perception of resulting injury or damage to plant
	• Inadequate management and supervisory attitudes
	• Low chance of detection due to inadequate supervision
	• Poor management or supervisory style
	• Poor accountability
	• Complacency caused by accident environments
	• Ineffective disciplinary procedures
	• Inadequate positive rewards for adopting approved work practices

By means of interviews and questionnaires applied to generic or specific sets of rules, the approach is designed to identify those set of rules and procedures within an organisation which, if not followed, could have the greatest potential impact on safety. Each set of rules or procedures is assessed using a checklist. A total of 48 questions have been created to map generic rule sets, as it would be impractical to complete a checklist for every safety rule of an organisation.

Examples from the checklist include:
- The rules does not always describe the correct way of working
- Supervision recognises that deviations from the rules are unavoidable
- The rules are not written in a simple language
- Rules commonly refer to other rules
- I have found better ways of doing my job than those given in the rules
- I often encounter situations where no prescribed actions are available.

A selection of the workforce is asked to rate the 'degree of agreement' with 48 statements with a score ranging between zero ('disagree') and six ('strongly agree'). The methods provide ways of analysing the answers and linking them to appropriate management strategies for minimising the potential for violations.

3   Conclusions

Human error is a normative concept. This means that human error statements are not descriptions of natural events or of state of affairs waiting to be observed, classified and counted. Describing something as a human error is a process of comparing actions to standard of correctness, i.e. ideal or existing rules defining correct execution for given tasks. Such comparisons are meaningful only within explicit or implicit frameworks of human performance, i.e. theories about the causes of human behaviour.

The human factor literature abounds with definitions of human error and its sub-categories (e.g. mistakes, violations), partly as a consequence of not recognising the normativity of the concept, i.e. rule-like statements instead of descriptive statements. This paper has pointed to areas of misunderstanding and difficulty related to current definitions, particularly the reliance on the concept of intention, as well as by showing the existence of a continuum between the commonly used notions of mistake and violation.

The second main task has been highlighting the major theories or models of human performance to see how they treat the concept of human error and how they classify and work with the concept. One major consequence that can be deduced is that errors classified by different systems cannot be compared directly as they make sense only within the particular models underlying the classification systems. Moreover, trying to learn by generalising from errors classified and counted even within a single system is problematic. In order to make assignments into a taxonomy, an explanation of the behaviour under the specific circumstances is required. This is allowed by the underlying theory of human behaviour. However, all serious theories of human performance recognise that behaviour is the result of several causes together with several conditions of executions, a complexity that generally becomes lost in the assignment to predefined static categories.

In a sense, those maintaining the concept of human error as being useless are claiming that we cannot build theories of human performance in technological systems (thereby improving their safety) from the concept of error (rather, we do the opposite), and that instead we should improve our theories by testing them on their empirical predictions rather than on their normative implications. This is a complex issue, which refers to the role of normativity in the natural sciences and, in this context, to what aspects of scientific theories are useful to 'test'. One thing is clear: although modern safety science has abandoned the obsession to precisely define, count and eliminate human error, it is still impossible to completely remove the need for the concept when working with safety-critical systems, especially for prospective evaluations and risk analyses. Yet, if we want to stay on productive grounds when it comes to human error, we have to go beyond folk views of the concept, the single major one being that errors are plain descriptions of events.

A set of suggestions to those engaged with the topic of human error can be derived as the conclusions of this paper:

1. *Specify the standard of correctness.* This follows the normative nature of the concept human error. Whenever something is defined as an error, a violation or an unsafe act the standard assumed should be mentioned.
2. *Differentiate between ideal versus actual standard.* Ideal standards are missions, principles of correct functioning, expected practices, etc., whereas actual standards are existing rules, procedures and instructions.
3. *Specify the point of view: internal versus external.* An internal point of view is the point of view of the agent, i.e. by looking at intentions and choice of goals. An external standard concentrates on the actual behaviour. As we have seen, this difference is relevant for defining errors.
4. *Indicate the model of behaviour adopted for the analysis.* The model of behaviour determines how the standards are applied to real situations, and what types of errors and behaviours are meaningful.
5. *Describe actions and conditions and distinguish them from the assumed causes.* This point stresses the importance of the context for action and reminds us not to forget it when classifying, as well as the importance of clearly differentiating between causes and manifestations.

These suggestions are more corollaries of the central theses in this paper than practical rules for error analysis. Nonetheless, it seems that keeping them in mind, or even using them as a checklist, will improve the accountability of one's analysis (whether defining, classifying or predicting something as human error), as all essential elements of error analysis will be made public. At the same time, when working with retrospective analyses, hindsight problems could be reduced as this process forces one to go through one's own assumptions and reasoning.

A sixth point could be added, which is more a terminological and methodological suggestion: *the concept of human errors is better suited for perspective and cognitive analyses*, while the concept of non-compliance (to safety rules and practices) appears more productive for working with organisational aspects.

Needless to say, these suggestions will not make the job easy. Standards of correct performance are seldom obvious or their application granted. Nor will these suggestions circumvent the problem of generalising the findings of incidents' analysis and error classifications. Further, there will always be several valid models of human behaviour and several associated classification systems.

References

Amalberti, R. (1992). Safety in Process-Control – An Operator-Centered Point-of-View. *Reliability Engineering & System Safety, 38,* 99-108.
Amalberti, R. (2001). The paradoxes of almost totally safe transportation systems. *Safety Science, 37,* 109-126.
Baddeley, A. D., & Hitch, G. (1974). Working memory. In G. H. Bower (Ed.), *The psychology of learning and motivation: Advances in research and theory* (Vol. 8, pp. 47-89). New York: Academic Press.
Bartlett, F. C. (1932). *Remembering: a study in experimental and social psychology.* Cambridge: Cambridge University Press.
Bove, T. (2002). *Development and Validation of a Human Error Management Taxonomy in Air Traffic Control.* PhD Risø National Laboratory, Roskilde.
CENELEC (1999). *Railway Applications: Systematic Allocation of Safety Integrity Requirements* (CENELEC Report No. prR009-004:1999 E). Brussels: CENELEC, European Committee for Electrotechnical Standardization.
Collier, S. (2000). *A Framework for Understanding Violations and Errors of Commission* (Rep. No. HWR-634). Halden: IFE.
Dougherty, E. (1995). Violation – Does HRA Need the Concept? *Reliability Engineering & System Safety, 47,* 131-136.
Hale, A. R. & Glendon, A. (1987). *Individual behaviour in the control of danger.* Amsterdam: Elsevier.
Hale, A. R., Heijer, T., & Koorneef, F. (2003). Management of safety rules: the case of railways. *Safety Science Monitor, 7.*
Heinrich, H. (1931). *Industrial accident prevention.* New York: McGraw-Hill
Hellriegel, D., Slocum Jr, J.W., & Woodman, R.W. (1998). *Organisational Behaviour.* 8th Edition. Ohio: South-Western College Publishing.
Helmreich, R. L., Klinect, J. R., & Wilhelm, J. A. (1999). Models of threat, error and CRM in flight operations. In (pp. 677-682). Austin, Texas, USA: The University of Texas at Austin.
Hidden, A. (1989). Investigation into the Clapham Junction railway accident. UK department of Transport report. London: Her Majesty's Stationery Office.
Hollnagel, E. & Woods, D. D. (1999). Cognitive systems engineering: New wine in new bottles. *International Journal of Human-Computer Studies, 51,* 339-356.
Hollnagel, E. (1993). *Human reliability analysis: context and control.* London: Academic Press.
Hollnagel, E. (1998). *Cognitive reliability and error analysis method: CREAM.* Oxford: Elsevier.
Hollywell, P. & Corrie, J. D. (2000). Reducing Violations on the Railways. What only Experience Can Teach. In London: Energy and Safety Division, IBC Global Conferences Limited.
HSE (1995). *Improving Compliance with Safety Procedures – Reducing Industrial Violations.* HSE Books.
Isaac, A., Shorrock S.T., Kennedy R., Kirwan B., Andersen H., and Bove T. (2001). *Technical Review of Human Performance models and Taxonomies of Human Error in ATM.* (HRS/HSP-002-REP-01), EATMP-EUROCONTROL, Brussels.

Isaac, A., Shorrock S.T., Kennedy R., Kirwan B., Andersen H., & Bove T. (2002). *Short Report on Human Performance Models and Taxonomies of Human Error in ATM (HERA)*, report HRS/HSP-022-REP-02, EATMP-EUROCONTROL, Brussels.

Kahn, R. L., Wolfe, D. M., Quinn, R. P., Snoek, J. D., & Rosenthal, R. A. (1964). *Organizational stress: Studies in role conflict and ambiguity*. New York: John Wiley.

Kelly, M. J. (1999). Preliminary Human Factors Guidelines for Traffic Management Centers. Federal Highway Administration, Report No. FHWA-JPO-99-042, National Technical Information Service, Springfield, VA.

Kirwan, B. (1994). *A guide to practical human reliability assessment*. London: Taylor & Francis.

Lawton, R. (1998). Not working to rule: Understanding procedural violations at work. *Safety Science, 28*, 77-95.

Leplat, J. (1993). Intention and Error – A Contribution to the Study of Responsibility. *European Review of Applied Psychology-Revue Europeenne de Psychologie Appliquee, 43*, 279-287.

Lewin, Kurt. (1935) *A Dynamic Theory of Personality. Selected Papers*. New York: McGraw Hill Custom Publishing.

Massaiu, S. & Kaarstad, M. (2007). Human and Organisational Barriers, Goal Conflicts and Violations: A Theoretical and Empirical Study. Report IFE/HR/E-2007/1330. Halden, Norway: Institute for Energy Technology.

Massaiu, S. (2006). Safety rules non-compliance in two Norwegian road traffic centres. Paper, European Safety and Reliability Conference ESREL 2006. Estoril, Portugal.

Mason, S. (1997). Procedural violations: causes, costs and cures. In F.Redmill & J. Rajan (Eds.), *Human Factors in Safety-Critical Systems* (pp. 287-318). Oxford: Butterworth-Heinemann.

Mason, S. (2000). Easy Way to Tackle Violations. In Energy and Safety Division, IBC Global Conferences Limited, London.

Minsky, M. A. (1975). A framework for representing knowledge. In P. Winston (Ed.), *The Psychology of Computer Vision*. New York: McGraw-Hill.

Norman, D. A. (1983). Position paper on human error. In *NATO Advanced Research Workshop on Human Error*, Bellagio, Italy.

Rasmussen, J., Duncan, K., & Leplat, J. (1987). *New Technology and human error*. John Wiley & Sons.

Reason, J., Manstead, A., Stradling, S., Baxter, J., & Campbell, K. (1990). Errors and Violations on the Roads: A Real Distinction? *Ergonomics* 33, 1315-1332.

Reason, J., Parker, D., & Lawton, R. (1998). Organizational controls and safety: The varieties of rule-related behaviour. *Journal of Occupational & Organizational Psychology, 71*, 289-304.

Reason, J. (1990). *Human Error*. Cambridge: Cambridge University Press.

Rochlin, G. (1993). Defining 'High Reliability Organizations in Practice: A Taxonomic Prologue. In Roberts, K. (Ed.), *New Challenges to Understanding Organizations*. New York, NY: Macmillan Publishing Company.

Ruiz Quintanilla, A. S. (1987). New Technology and Human Error: Social and Organizational Factors. In J. Rasmussen, K. Duncan, & J. Leplat (Eds.), *New Technology and Human Error* (pp. 125-128). Chichester, UK: John Wiley & Sons.

Rumelhart, D. E. (1975). Notes on a schema for stories. In D. Bobrow & A. Collins (Eds.), *Representation and Understanding: Studies in Cognitive Science*. New York: Academin Press.

Schmidt, R. A. (1975). A schema theory of discrete motor skill learning. *Psychological Review, 82*, 225-260.

Senders, J. W. & Moray, N. P. (1991). *Human Error: Cause, Prediction, and reduction*. Hillsdale, New Jersey: Lawrence Erlbaum Associates.

Skitka, L. J., Mosier, K., & Burdick, M. D. (2000). Accountability and automation bias. *International Journal of Human-Computer Studies, 52*, 701-717.

Skitka, L. J., Mosier, K. L., & Burdick, M. (1999). Does automation bias decision-making? *International Journal of Human-Computer Studies, 51*, 991-1006.

Swain, A. (1963). A method for performing a human factors reliability analysis, Monograph SCR-685, Sandia National Laboratories, Albuquerque, US.

Swain A. & Guttmann, H. (1983). *Handbook of Human Reliability Analysis with Emphasis on Nuclear Power Plant Applications.* NUREG/CR-1278, Nuclear Regulatory Commission, U.S.

Taylor, D. H. (1987). The Hermeneutics of Accidents and Safety. In J. Rasmussen, K. Duncan, & J. Leplat (Eds.), *New Technology and Human Error* (pp. 31-41). Chichester, UK: John Wiley & Sons.

Watson, J. B. (1913). Psychology as the behaviorist views it. *Psychological Review,* 20, pp. 158-177.

Wickens, C. D. (1992). *Engineering psychology and human performance.* New York: HarperCollins.

Wilde, G. J. S. (1974). Wirkung und Nutzen von Verkehrssicherheitskampagnen: Ergebnisse und Forderungen – ein Überblick. *Zeitschrift für Verkehrssicherheit,* 20, 227-238.

Woods, D. D., Johannesen, L. J., Cook, R. I., & Sarter, N. B. (1994). *Behind Human Error: Cognitive Systems, Computers, and Hindsight.* CSERIAC Program Office.

Woods, D. D. & Cook, R. I. (2003). Mistaking error. In M. J. Hatlie & B. J. Youngberg (Eds.), *Patient safety handbook,* Jones and Bartlett.

Woods, D. D. (1986). Paradigms for Intelligent Decision Support. In E.Hollnagel, G. Mancini, & D. D. Woods (Eds.), *Intelligent decision support in process environments* (pp. 153-173). Berlin: Springer.

Woods, D. D. (1988). Commentary: Cognitive Engineering in Complex and Dynamic Worlds. In E.Hollnagel, G. Mancini, & D. D. Woods (Eds.), *Cognitive engineering in complex dynamic worlds* (pp. 115-129). London; San Diego: Academic Press.

2

A Formative Analysis of the Train Driving Task
Synve Røine

This paper presents a formative analysis of the train driving task. The train driving task is embedded in the railway domain which can be described as a complex socio-technical system. Such systems place new demands on the analytical tools used in work analysis. The formative approach is thought to deal with these demands. The framework applied draws on Cognitive Work Analysis. It describes the task in terms of the behaviour shaping constraints embedded in five conceptual distinctions: work domain, control tasks, strategies, social organisation and cooperation, and worker competencies. The focus is on describing the task in broad terms, and showing the interactions between the different conceptual distinctions and how they shape the task. The article draws upon multiple data sources. Observations were made on two local traffic routes in Norway. Interviews with drivers and safety personnel, and studies of secondary data were also done. The framework applied here proved promising in showing how the train driving task is shaped through the constraints on the different levels. It also showed how the drivers dealt with context-conditioned variability. Despite problems with device-dependence and level of detail, the framework can provide a starting point for further studies. It can also provide a means of displaying data from formative analyses in other domains in order to demonstrate how constraints shape behaviour.

1 Introduction

The rapid technological changes that have taken place during the last fifty years have presented us with increasingly complex socio-technical systems (Perrow, 1999; Vicente, 1999). Socio-technical systems are systems which are composed of technical, psychological and social elements. They can be described as complex if rated highly on some of the following dimensions: large problem space; social; heterogeneous perspectives; distributed, dynamic; potentially hazards; many coupled subsystems; automated, uncertain data; mediated action via computers; and disturbance management (Vicente, 1999; Woods, 1988). Well-known examples are industrial control centres, air traffic control, the bridge on a ship, emergency rooms, nuclear power plants, etc. Complex socio-technical systems represent progress in the way that they have given us the possibility to control and coordinate large forces and resources. At the same time, there are potentially high hazards associated with the complexity of some of these systems and the impact of failure is more severe than with simpler systems. The increased risk and prevalence of accidents with catastrophic consequences are tell-tale signs of this (Casey, 1998; Perrow, 1999; Vicente, 1999). There is a need to increase our understanding of these systems in order to improve safety, productivity and health. This is the rationale behind the research area of cognitive engineering, which is concerned with the analysis, design and evaluation of complex socio-technical systems (Vicente, 1999; Woods & Hollnagel, 2006).

1.1 Railways as a complex socio-technical system

Wilson and Norris (2006; 2005) describe railways as a complex socio-technical system. Many different stakeholders cooperate and act together with technology in order to transport passengers and goods safely and efficiently. The system is also described as social, as many different people are cooperating in many different subsystems. The different actors (e.g. drivers, conductors, traffic control centre operators, maintenance workers, and station employees) are also located in different places and the system can thus be described as distributed (Wilson et al. 2001). Kecklund (2001) described the train driving task as being in an information vacuum, which suggests uncertain data. The train system is also characterised by automated parts of the work (Kecklund, 2001; NSB AS, 2007). Railroads are commonly considered a safe transport domain (Wilson & Norris, 2006). In Norway the number of deaths caused by railway accidents has been relatively stable for the last 30 years. In 2005, 3 people were seriously injured and 3 killed (Norwegian National Railway Administration, 2006). By comparison, 997 people were seriously injured and 224 were killed in traffic in the same year (Statistics Norway, 2006). However, the 'Åsta accident' in Norway and recent high-profile accidents, such as the Cumbria derailment in England on 24 February 2008 and the train crash in France near the Luxembourg boarder on 11 October 2006, have contributed to an increased focus on rail safety. The rail domain is now, after years as a relatively slow changing domain, undergoing rapid technological change. Wilson and Norris (2006) predict that this trend will persist in the years to come as a result of heavier investments due to increased passenger load and environmental and transport economy considerations. This is also the case for the local traffic areas around Oslo, where new rolling stock is operating on the outer local route and the equipment on the inner local routes is due to be replaced within a five-year period (NSB, personal communications, 2007).

1.2 Cognitive Engineering: New Demands on Analytical Frameworks

The train driving task is embedded in the complex socio-technical system of railways. The study of work in these complex systems imposes new demands on how to study work, and it is important to find a method of analysis that takes this into consideration. Rasmussen (1997) makes a distinction between the normative, the descriptive and formative approaches to work analysis. This provides a useful taxonomy for comparing a number of different techniques with regards to the fundamental assumptions and problems of applying them to complex socio-technical systems. A discussion of these different techniques will follow after the descriptions of the three different approaches.

1.3 Normative approach

The normative approach divides the work into different tasks and subtasks, which in turn describe the different steps needed to achieve a specified goal, and describe work in a 'best way' manner. Tayloristic work methods, the Goals, Operators, Methods and Selection rules model (GOMS), and task analysis are all examples of normative apporaches (Vicente, 1999). The use of these methods is prevalent in Human Factors (Shepherd & Marshall, 2005). A task analysis is, for example, a detailed way of optimising the human element within a system in a systematically, open manner which can be subject to careful scrutiny (Kirwan & Ainsworth, 1992). The normative approach to work analysis is efficient because it identifies what needs to be done and how, and in this way provides us with a simple model of work (Vicente, 1999).

Railways have only recently obtained extensive attention from the human factors community (Wilson & Norris 2006). Traffic control centres is one of the areas that has received a lot of attention in the rail domain. Published studies of the train driving task are, on the other hand, still relatively few but Hierarchical Task Analyses (HTA) of the train driving task have been done by Shepherd & Marshall (2005) and Rizzo, Pozzi, Save & Sujan (2005).

1.3.1 Evaluation of the Normative Approach

Despite its prevalence in human factors, the normative approach encounters a few problems when applied to complex socio-technical systems. The underlying assumptions of the normative approach are that the initial conditions are known, that influences can be accounted for and that goals can be clearly stated (Vicente, 1999). Conditions need to be fixed in order to define a task (Leplat, 1989), and initial conditions are often unknown in complex socio-technical systems. In addition, they are subject to external influences, making them open systems. This means that there are patterns of disturbances in the system which will rarely be repeated because it would require different types of action. This is referred to as context-conditioned variability, a term adopted from motor control literature (Vicente, 1999). The incapacity to specify the variance in the context causes a discrepancy between the task realised in context and the planned task (Gautherau & Hollnagel, 2005). Increasing complexity makes it more difficult to accurately predict tasks as well as dividing tasks into subtasks (Leplat, 1994). Clearly identifiable goals are, in addition, not always possible to find, and in special circumstances, for example, associated with emergencies and accidents, clear goals can be hard to identify (Hollnagel, 2004). The normative approach will therefore suffer from some limitations when applied to the study of work in complex socio-technical systems.

1.4 Descriptive Approach

One of the main arguments against the normative approach is that it does not portray work in a realistic way. The descriptive approach offers an alternative to this as it tries to *'describe how a system actually behaves in practice'* (Vicente, 1999, p. 61). It is descriptive in the way that it tries to document and understand current practices through field studies and naturalistic observation. Studies of situated action (Suchman, 1987), naturalistic decision making (Klein, 1995; Zsambok & Klein, 1997), activity theory (Bødker, 1991; Nardi, 1997), and distributed cognition (Hutchins, 1995), are all examples of descriptive approaches to work analysis (Vicente, 1999). The findings from the different descriptive approaches have two common themes. The first theme is the focus on the study of work in representative or naturalistic settings. The picture that emerges from studying both the work in situ and what workers actually do is quite different from the one presented by the normative approaches. It shows how humans make things work in real life and draws a picture of workers as adaptive and ingenious in their everyday practices. The second theme that appear in these studies is the converging characterisation of human work, as they show the importance of context conditioned variability, a strong social component, external mental processing in relation to the development and use of tools, time pressures, and that work is shaped by historical-cultural factors (Vicente, 1999). There have been few descriptive studies of the train driving task, but one example is a case study of the Italian railway system in the investigation of distributed cognition (Eurocontrol, 2003).

1.4.1 Evaluation of the Descriptive Approach

The strength of the approach is how it portrays work as it is. Still, studies from social science, activity theory, francophone ergonomics, and human factors all show a weakness in relation to linking the findings to the design of new practices and artefacts. The fact that the work situation studied is representative of the actual work situation does not ensure relevance to intervention or design (Hoc, 2001). Design ideas are often based on the study of current practice (Vicente, 1999). A new artefact will alter how work is done when it is introduced into the domain, and the result is new work practices. These practices are usually not adequately supported by the novelty and can result in new problems. Hence, a new device intended to support work may actually transform the work and present entirely new challenges (Carroll, Kellogg & Rosson, 1991; Sarter & Woods, 2000; Dekker & Woods, 2002; Woods, 1998). This was seen, for example, in an evaluation of the first ten years of the Automatic Train Control (ATC) system in Sweden, where the drivers altered their driving style to fit the ATC system (Olhsson, 1990 as cited in Kecklund, 2001). The task-artefact cycle introduced by Carroll et al. (1991) depicts this interdependence between current practice and design and shows the circular nature of technological innovation. Thus, the descriptive analysis becomes outdated the moment we use it to alter an artefact or practice (Vicente, 1999).

1.5 Formative Approach

The focus of the formative approach is *'identifying requirements – both technological and organisational – that need to be satisfied if a device is going to support work efficiently'* (Vicente, 1999, p.61). The formative approach is concerned with modelling behaviour shaping constraints, defining constraints as either relationships between or limits on behaviour (Vicente, 1999). The idea is that human behaviour is governed by constraints, both associated with the system and within the human, and that these constraints must be respected for successful work performance (Rasmussen, 1990). The focus is on specifying the requirements rather than the prescribed steps associated with a task. This enables the formative approach to deal with the context-conditioned variability of complex socio-technical systems. The formative approach strives to escape the task-artefact cycle by describing the task without referring to the existing technical solutions, which is also referred to as device-independence. This approach to work analysis can provide a solution to the problems associated with the normative and descriptive approaches (Rasmussen, 1997; Vicente, 1999). Finally, it is important to remember that the formative approach does not necessarily exclude normative or descriptive techniques (Vicente, 1999).

Rasmussen (1990; 1997) and Rasmussen, Pejtersen, & Schmidt (1990) suggest a version of a formative work analysis. Vicente (1999) elaborates on this in his Cognitive Work Analysis (CWA) framework. CWA is a formative framework based on the analysis of work in terms of five conceptual distinctions, and this framework is used to model how constraints shape work. Each of the conceptual structures are further linked to a set of modelling tools and to different systems interventions, and the aim of CWA is to develop new work practices (Vicente, 1999). This paper will build on this approach and apply certain aspects of it.

There have not been many studies of the train driving task using a formative approach, but Jansson, Olsson & Erlandsson (2006) used CWA to investigate the train driving task, with the aim of improving the Automatic Train Control (ATC) interface. Biemans (2006) also used the CWA framework in a study of train driving.

1.5.1 Applying the formative approach

The formative approach deals with many of the challenges in the study of work in complex socio-technical systems. CWA is very resource demanding, and Vicente (1999) sees this as the main weakness of the approach. He does, however, open up for the possibility of focusing on a subset of the analysis in order to deal with specific problems. This was done in a study by Jansson et al. (2006), who performed an analysis of a work domain which resulted in the redesign of the ATC interface. A focus on a specific problem and/or subset, however, implies sacrificing the more overall picture of the train driving task, and this can be problematic in two ways. First, the key in Cognitive Engineering(CE) is to understand the various layers in the complex socio-technical system and more importantly the interactions among them (Vicente, 1999). This means that it is valuable to strive to achieve a description of the train driving task in terms of all the various layers and their interactions. Second, the focus on specific problems entails other interesting features being overlooked. I believe that it is premature to decide such a specific problem or feature at this point, because we risk focussing on the wrong problem.

Thus, the CWA framework should not be used in its complete form. In this paper, I will take a broad perspective, and describe the various layers in the complex socio-technical system of the railway and the interactions among these layers. I also hope to reduce the risk of overlooking important aspects by providing an overview of the train driving task.

1.6 The Formative Approach in this Paper

This paper will use the conceptual distinctions defined in CWA (Vicente, 1999) and incorporate them into a framework developed for this study. The aim is to show how the constraints associated with the different conceptual distinctions shape the task, and the hierarchical interactions among the different levels. The framework will be described in the following.

The work domain is analysed first as it outlines the boundaries and the extent of the system being controlled. It is defined as *'the system being controlled, independent of any worker, automation, task, goal, or interface'* (Vicente, 1999, p. 113). The next four conceptual distinctions are analysed together. The different control tasks pose different demands or constraints on the strategies chosen, which in turn affects the social organisation and cooperation, and also the worker competencies required. It is these behaviour shaping constraints that the formative analysis strives to reveal, and the four remaining conceptual distinctions will therefore be analysed together following analysis of the different control tasks.

The control tasks focus on identifying what needs to be done, meaning the goals to be achieved. They can be defined as *'the goals that need to be achieved, independently of how they are to be achieved or by whom'* (Vicente, 1999, p. 113). These goals can be differentiated in terms of constituting the overall goal or control task and more specific subtasks. Control tasks are separated into two categories because they represent slightly different levels of analysis.

Strategies are *'the generative mechanisms by which particular control tasks can be achieved, independently of who is executing them'* (Vicente, 1999, p. 114). They are investigated in relation to the defined control tasks. This conceptual distinction is also divided in two, according to different levels of analysis. The higher order strategies describe the strategies for action in general terms, while the lower order strategies are closer to the strategies actually carried out by the system.

Social organisation and cooperation deal with *'the relationships between the actors, whether they be workers or automation'* (Vicente, 1999, p. 114). This allocation of tasks and responsibilities is analysed in relation to their respective strategies.

Worker competencies are the *'constraints associated with the workers themselves'* (Vicente, 1999, p. 114). The worker competencies discussed in this paper are analysed from the driver's point of view. I therefore only refer to drivers' competencies when commenting on worker competencies in this paper. These are the human capabilities and limitations, and particular competencies, which need to be exhibited if they are to function efficiently (Vicente, 1999). I have followed this division between human capabilities and competencies in the analysis of worker competencies. Human capabilities point to the possibilities and limitations of human functioning. Here, competencies will refer to more complex skills and knowledge acquired through learning. I have further organised the competencies in terms of different types: technical competency (skills and knowledge concerning the train and the equipment), procedural competency (knowledge about the rules, routines and procedures), environmental competency (knowledge about the effect of weather conditions, the characteristics of the route, etc.), and social competency (knowledge about the behaviour of passengers). Studies show that these competencies can be formal, as learned through education, or more experience based (see e.g. Biemans, 2006; Kecklund, 2001), and I will also make use of this distinction.

This paper will use the formative analysis outlined above to investigate the train driving task.

1.7 Study Objectives

The starting point for Cognitive Engineering is the world of practice, demand for interventions, and tools (Maramas & Nathanael, 2005). The train driving task is embedded in a complex socio-technical system of railways, which for the purpose of this paper constitutes the world of practice. The demand for intervention is due to the increased focus on rail safety and the rapid technological change in the domain (Wilson & Norris, 2006). The characteristics of railways as a complex socio-technical system also make new demands on the tools we apply. The formative approach to work analysis described here provides a viable alternative for investigating the train driving task, and hence the first aim of this paper is:

1) To make a formative analysis of the train driving task and describe it in terms of the behaviour shaping constraints in the five conceptual distinctions.

Cognitive Engineering needs to refine its methods and models (Vicente, 1998). The version of a formative analysis suggested in this paper borrows aspects from the

established CWA framework (Vicente, 1999) and adds new ones. It is necessary to evaluate the framework presented in this paper in order to assess its contribution to the field. The second aim of the paper is therefore;
2)   To evaluate the formative analysis framework applied in this paper.

2   Methods

There is a need to draw on multiple methods when investigating complex socio-technical systems (Hollnagel & Woods, 2005; Nardi, 1997; Vicente, 1999). The methods used for this study were observations, interviews and secondary data sources.

2.1   Observations

Onboard observations were made on the Norwegian State Railways' (NSB AS) local routes during ordinary operations in February 2007. Two types of train and routes were selected for study. Both are local routes with a heavy load of passengers and many stops. The inner local route Skøyen–Ski has the highest frequency of stops of the two and is operated by train type 69. Trains of this type were produced in Norway

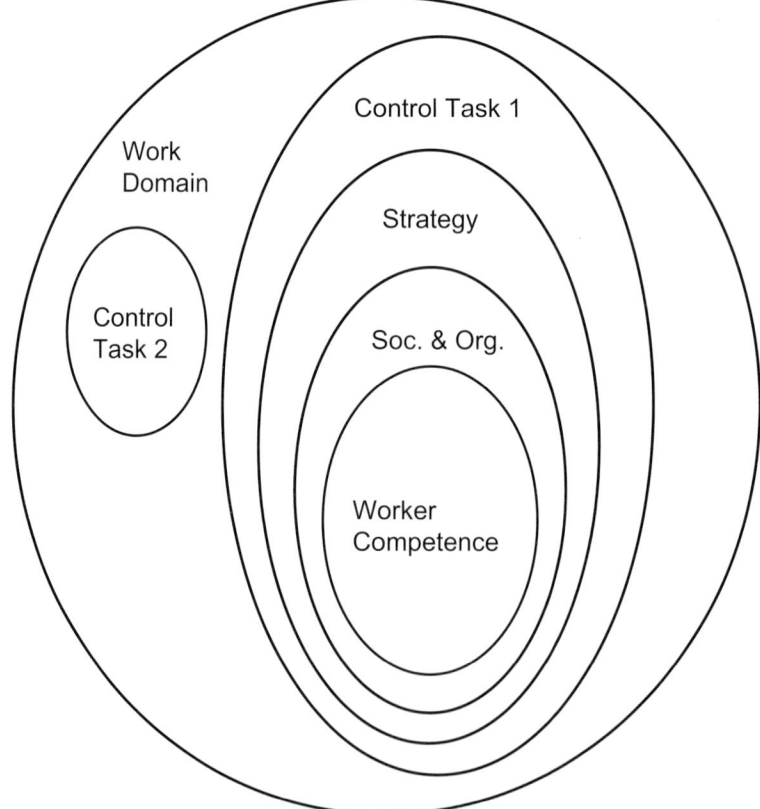

Figure 2.1 Nesting of the conceptual distinctions. Adapted from Vivente (1999) p. 116.

between 1970 and 1993 but have been upgraded with newer safety equipment, e.g. ATC and new telephone systems. The outer local route, Spikkestad–Moss, has fewer stops and is operated by train type 72, a newer model produced in Italy between 2002 and 2004.

The participants comprised nine experienced train drivers who participated voluntarily. They were presented with information about the study, and informed about their rights as participants before signing consent forms.

2.1.1 Direct observation

The three direct observations were made in the drivers' cabs of types 72 and 69, with five different drivers, and during normal operations. The observations were made both with a safety official present and when the drivers were one their own. I had the opportunity to ask questions while travelling along the route. The observations were made using a hand-held DVD camera, notes and a photo camera.

The cognitive walk-through technique was used during the direct observations (see e.g. Kirwan & Ainsworth, 1992), and the train drivers were asked to comment on the actions they were performing while being filmed. This allowed a record to be kept of both the non-observable and observable actions that were taking place as well as the details of the procedures.

2.1.2 Indirect observation using video recording

The direct observations were used to form a basis for the two indirect observations by providing information about how and where to place the video equipment. One trip with the type 69 train and one with the type 72 train was conducted, with two different drivers on each trip. Two hand-held DV cameras were taped to the interior of the train cab. They were positioned to show the actions of the driver and the route ahead. The camera directed towards the driver was equipped with a wide-angle lens in order to secure a view of the whole cab environment. The two perspectives were later synchronised and merged into a single screen, allowing for the analysis of the drivers' actions in relation to the environment outside the train.

2.1.3 Operational Sequence Diagrams

The video material from the indirect observations was analysed by means of a task analysis. The analysis chosen was an Operational Sequence Diagram (OSD), which shows the sequence of control movements and information on collective activities involved in a task. This type of analysis can be conducted in several ways and the type chosen for this study was a partitioned OSD, which focuses on specific criteria (Kirwan & Ainsworth, 1992). In this case, the criteria investigated were the operations associated with a station area (approaching and leaving) and being on the line between stations. These criteria were based on the direct observations (see Appendix 1 for a sample of two OSDs from a station area). The direct observations and interviews also informed the development of the OSDs.

2.2 Interviews

Interviews were used throughout the study during the direct observations as well as separately. The safety officials and train drivers were interviewed using unstructured interviews in the initial phases. In later phases, three semi-structured interviews were conducted. The participants were train drivers with an average of 22 years of experience. The aim of holding the interviews was gain insight into the train drivers' own stories about the different challenges associated with their work. They were also encouraged to elaborate on specific episodes when they had been particularly challenged or under a lot of stress.

2.3 Secondary Data Sources

Several secondary data sources were used and incorporated in the analysis throughout. These included annual reports and official laws and regulations which are publicly available on the Norwegian Railways Inspectorates web pages (FOR 2002-01-29 nr. 122; FOR 2002-01-29 nr. 123; FOR 2002-12-18 nr. 1678; FOR 2002-12-18 nr. 1679; FOR 2001-12-04 nr. 1335; FOR 2001-12-04 nr. 1336). Publicly available annual reports from the Norwegian National Rail Administration (NNRA) and the Norwegian State Railways (NSB AS), in addition to internal documents from NSB such as technical and safety documents were also used. Extensive use was made of NSB's traffic safety manual (NSB AS, Persontog drift, 2007), which includes rules, regulations and procedures for passenger traffic. Accident reports from the Norwegian Accident Investigation Board (AIBN) were also examined.

3 Results

The results will be described in terms of the five conceptual distinctions: work domain, control tasks, strategies, social organisation and cooperation, and worker competencies. The work domain (3.1) is presented first before the next four (3.2), which are presented together to demonstrate how the constraints of higher conceptual distinctions influence lower ones.

3.1 Work Domain
3.1.1 Stakeholders

The Norwegian railway system comprises several organisations and stakeholders. The Norwegian Railway Inspectorate (NRI) is the control and supervisory authority for all stakeholders (Norwegian Railway Inspectorate, n.d.). They are responsible for ensuring that rail traffic is operated safely and appropriately in the best interest of the different stakeholders. The NRI is also responsible for drawing up regulations, awarding licenses, and approving infrastructure and rolling stock. The Norwegian National Railway Administration (NNRA) is responsible for the infrastructure. They control the traffic via their traffic control centres, and are responsible for the maintenance of tracks, signs, signals, and stations. There are ten traffic operators including cargo and passenger traffic (Norwegian National Railway Administration, n.d.). The biggest company dealing with passenger traffic is the Norwegian State Railways (NSB), which had 49 million passengers in 2006 (NSB, n.d.).

3.1.2 Infrastructure and traffic conditions

The transportation of passengers is divided into different routes, and two local routes around the Oslo area were investigated for this study. The Moss–Spikkestad route is an outer local route with 21 stops and a duration of 1 hour and 35 minutes, and the Ski–Kolbotn route is an inner-local route with 24 stops and a duration of 40 minutes.

Both lines are operated on the same double tracks (one in each direction). At Ski station the inner local train heads back while the outer local train continues to Moss. The line is one of the most heavily trafficked lines in Norway as it is not only operated by local traffic but also by cargo and intercity trains, in addition to being the international routes from the Continent to Oslo, via Sweden. The Norwegian National Railway Administration's control centre in Oslo is responsible for controlling traffic by operating the lights signals along the tracks (hereafter referred to as the signal system).

3.1.3 Weather

The weather influences the railway system by altering the trains' driving performance and the drivers' conditions for observation. Ice can make the platforms slippery during the winter, and snow can impair the drivers' view. Cold weather can also affect the power of the brakes and this has especially been a problem for type 72 trains. Winters in Norway are characterised by short dark days. The darkness increases the visibility of the signals, but it also decreases the visibility of disturbances along the line, such as people and animals. The problem is quite the reverse during the summer season: the bright sunshine can make the light signals more difficult to see. In the autumn, wet leaves can create a problem because they make the tracks very slippery and impair braking.

3.2 Control Tasks, Strategies, Social Organisation and Cooperation, and Worker Competencies

The constraints related to the aforementioned four conceptual distinctions will be presented together in this section. The control tasks are divided into the overall task and subtasks. This is also the case for strategies, which are divided into higher order strategies and lower order strategies. Social organisation and cooperation is kept as a single category. Worker competencies are described as reflecting both human capabilities and competencies. Human capabilities reflect the possibilities and limitations of human functioning, while competencies point to more complex skills acquired either through formal education or experience.

3.2.1 Overall Task and Presentation of Results

The overall task or goal of the local traffic is to *safely, efficiently and comfortably transport passengers from A to B on time*. This overall goal can further be broken down to three subtasks: *travel between stations in order to reach stations on time* (on the line) (section 3.2.2); *stop at station to let people on/off* (approaching station) (3.2.3); *and leave station on time* (leaving station) (3.2.4). These subtasks are related to the different phases of driving in the study. The order of the presentation of these subtasks does not reflect any difference in their significance.

Each subtask will be described separately in terms of strategies, social organisation and cooperation, and worker competencies. The emphasis is on how the different layers affect each other and this is achieved by describing the subtasks horizontally. The structure of the results is shown in Table 2.1.

Table 2.1 Organisation of conceptual distinctions in the analysis.

Control task			Strategies		Sos. and Org	Worker competencies
Overall task	Subtask		Higher order strategies	Lower order Strategies		
	3.2.2	On the line	→	→	→	→
	3.2.3	Approaching station	→	→	→	→
	3.2.4	Leaving station	→	→	→	→

For the following section, readers may find it helpful to look briefly at the tables before and after reading the details of the different subtasks.

3.2.2 Out on the line

The first subtask described is to ensure safe, comfortable and efficient travel between the stations in order to reach the stations on time. This can further be divided into three higher order strategies: 1) keep the right speed, 2) maintain safety of people and equipment, and 3) make sure equipment works adequately.

For an overview see Table 2.2.

1) Keep the Right Speed
Higher order strategy
The first higher order strategy is keeping the right speed in relation to safety, the timetable, economic considerations and comfort. This is important in order to reach the stations on time and not cause delays on a heavily trafficked line.

Lower order strategies
The lower order strategy is to monitor the speed, and adjust it in relation to a number of factors: the timetable, the weather, traffic, technical status, and disturbances on the tracks.

Social organisation and cooperation
There are several actors on the level of social organisation and cooperation. The signals, signs and speedometer, together with the Automatic Train Control (ATC), are the main instruments to maintain speed within safety limits. The ATC system was developed to ensure that trains do not pass stop signals, a situation called signals passed at danger (SPAD). The ATC will first give an audible warning, then if necessary brake and stop the train if the train is in danger of passing a stop signal. The ATC also helps the driver to keep the right speed by calculating the train's braking curve. Norwegian railways do not have full ATC coverage. A distinction is made between areas without ATC, Full Automatic Train Control (FATC), and Partial Automatic Train Control (DATC). The maximum speed in FATC areas is kept under surveillance at all times, in contrast to areas with DATC where only the main signals and the pre-signals are controlled. This means that the ATC in FATC areas provides the driver with information about the current speed and the target speed, while in the DATC areas it only serves as a safeguard against SPADs. The routes studied have areas of both FATC and DATC, and these are marked with signs to inform the drivers about the change in coverage. This difference in coverage means that task allocation between the driver and the ATC in terms of speed maintenance and stop signals is not static during travel between stations.

Speed is also adjusted in accordance with the route book, which shows the driver the stations on the route and the departure time for each station on the route. Together with the route book, the time is important in determining what the right speed is, because an increase in speed can help a driver recover time if the train is running a little behind schedule. In addition to keeping to the timetable, drivers are required to take considerations of economy into their driving as well. For example, it is uneconomical to drive fast if this only results in arriving at a station too early.

The traffic control centre is more directly involved when there are cases of special incidents in the traffic or, more commonly, when signals are faulty. Every train is equipped with a phone link to the traffic control centre, and the drivers are contacted in cases of signal error or traffic incidents. The drivers can also contact the traffic control centre, and this is common in the case of faulty signals. It is specified in the traffic safety manual that drivers are required to contact the control centre whenever they have to wait too long at a stop signal. The train can then pass a stop signal if permission is received from the traffic control centre, which is done in order to keep the traffic flowing. The other actors involved in speed regulation, such as the ATC, the signs and the signal systems, are all fixed and do not take weather and technical status into consideration, as it is the driver's responsibility to do this.

Worker competencies
Technical, procedural and environmental competence is important at the level of worker competencies. The act of regulating speed is a skill acquired during drivers training'. To know how to regulate speed safely and efficiently entails the knowledge of how to incorporate procedural knowledge from education, rules and regulations with experience-based knowledge about the environmental conditions and technical status.

2) Maintain Safety of People and Equipment
Higher order strategy
The second higher order strategy is to keep people and equipment safe. People, animals and objects may find their way onto the tracks and lead to dangerous situations.

Lower order strategies
The lower order strategies are to monitor the tracks and the environment for disturbances, and be prepared to brake.

Social organisation and cooperation
Tasks are carried out through a combination of people and artefacts. The signals and signs, and the ATC are instrumental in securing that no other trains are in conflicting positions. This task allocation actually changes as the train drives through areas with either FATC or DATC. The driver also has to monitor the environment for other disturbances on the tracks. The safety brake system (SIFA) is an automatic system which is designed to ensure that the driver is awake and alert. The system makes a sound at regular intervals, and stops the train if the driver does not respond to the sound by stepping on a pedal. If there are disturbances on the tracks the only alternatives are to try stop and/or signal. The train's long braking distance does not allow it to stop rapidly, so the driver has to detect any disturbances early in order to have a chance of avoiding impact. This means that the driver needs to be prepared to brake at all times, but an auditory signal (horn) is often the only real alternative. The standard procedure is to brake and use the horn simultaneously. Especially risky places are equipped with signs that tell the driver to give the auditory 'train is coming' warning. The social organisation and cooperation is characterised by the involvement of many actors. It also shows that there are limited means available to prevent impact with objects or persons while on the route.

Worker competencies
The most relevant human capability in this context is attention. A train's high speed and slow response time makes the driver's response time more important. The driver also needs to be aware of whether s/he is in a FATC or a DATC area, since this changes the nature of their task. The procedures and rules associated with this task are present in drivers' formal procedural competencies. Drivers' environmental competencies are also relevant, and this can be knowledge about especially risky places, bad

weather conditions, and risky times of day and year. Examples are: A nearby school that increases the risk of children being on the platform, snow in the winter, Friday nights when people are on their way home from a party, or springtime when there are more tractors working closer to the tracks.

3) Make Sure Equipment Works Adequately
Higher order strategy
The third and final higher order strategy associated with the subtask of being on the line is to make sure that the train and equipment work adequately.

Lower order strategy
The high order strategy is carried out by the lower order strategy of monitoring the equipment inside and outside the cab, and this strategy can be found in all of the three subtasks.

Social organisation and cooperation
The social organisation and cooperation is distributed between the driver and equipment. Lamps and the IDU (input/display unit) screen monitor the condition of the various parts of the technical system on the train. The IDU screen is a touch screen linked to a computer which monitors the status of the train at all times, and is used to diagnose the problem in the case of 'deviations' (i.e. all deviations from the prescribed condition). The IDU screen is present in the newer trains such as the type 72 but not in type 69 trains. The task of diagnosing error in the type 69 is divided between the warning lights in the cab and the driver. If there is a problem, the driver must walk around in the train and carry out visual checks. All trains are equipped with phones connected to a technical backup centre known as DROPS. The drivers can contact DROPS if they need additional assistance in solving technical problems.

Worker competencies
The worker competencies associated with a lower order strategy are important. Drivers need extensive technical knowledge of how the train works and this is provided by their training. Experience also gives them schemas of what can be wrong and how to fix it. The information from the sounds, vibrations and mirrors can be valuable for the experienced driver in assessing the technical condition. The tight schedule on the local routes makes this task of fixing technical errors important for the overall traffic conditions. It is the drivers' responsibility to resolve errors fast if they occur. All the drivers reported that the types 69 and 72 trains were regularly subject to errors. Especially, the new type 72 was reported to need a lot of attention, and even the IDU screen contributed to problems by reporting non-existent errors on a regular basis.

Table 2.2 On the line - the subtask of safely, comfortably, and efficiently travel between the stations.

Control task		Strategies		Sos. and Org	Worker competencies
Overall task	Subtask	Higher order strategies	Lower order Strategies		
Safely, efficiently, and comfortably transport passengers from a-b on time	Safely, comfortably, and efficiently travel between the stations in order to reach the stations on time	1) keep right speed - safety - timetable - econmic - considerations - comfort	Monitor speed and adjust according to - signals - timetable - weather - traffic - technical status - disturbances on the tracks	Driver Signals ATC Route book Train Speedometer Phones – Traffic – Control Centre	Human capabilities – Attention Competencies – Technical – Procedural – Environmental
		2) maintain safety for people/ equipment	Monitor tracks and the environment for disturbances Be prepared to brake and signal	Driver Mirror Signals & Signs Horn	Human capabilities – Attention Competencies – Technical – Procedural – Environmental – Social
		3) Make sure train/equipment workers adequately	Monitor train/equipment	Driver IDU Lamps Mirrors DROPS	Human capabilities – Attention Competencies – Technical – Procedural – Environmental

3.2.3 Approaching station

The subtask of safely, comfortably, efficiently stopping at a station to let people on/off can be divided into seven higher order strategies: 1) Detect the right station, 2) Inform passengers about the next station, 3) Maintain safety of people and equipment, 4) Stop smoothly and safely, 5) Maintain safety of people on station, 6) Let people on and off, and 7) Make sure that the train and equipment are functioning adequately.

For an overview see Table 2.3.

1) Detect the right station
Higher order strategy
The first higher order strategy is to detect the right station, and this is important because the train does not always stop at each station.

Lower order strategies
The lower order strategy associated with detecting the right station is monitoring the environment both inside and outside the driver's cab.

Social organisation and cooperation
The social organisation and cooperation divides the lower order strategy between the driver and equipment. The route book shows the driver all of the stations on the route and the departure time for each station. Signs along the tracks also inform the driver that the train is approaching a station, and in type 72 trains this task is further supported by the automatic station announcer. The announcer is GPS (global positioning system) controlled and although it is primarily for the benefit of the passengers it is also heard by the driver.

Worker competencies
The human capabilities needed to fulfil the strategies are attention to the information presented. The competencies required are the procedural knowledge from drivers' training in the meaning of signs and signals. However, experienced drivers report that they know the route so well that at all times they know which station is next on the route. This emphasises the importance of the informal competencies acquired through experience. The primary source of information for these informal competencies lies in the environment, namely in the different topographic features along the track.

2) Inform passengers about the next station

Higher order strategy
The second higher order strategy is to inform the passengers of the next station, so they can get off at the right station, fast and efficiently. Rapid stops at each station are required to keep to the timetable.

Lower order strategies
The lower order strategy is to announce the next station for the passengers.

Social organisation and cooperation
The social organisation and cooperation is differently solved in the two train types. In type 72 trains, the automatic station announcer responsible and the information is also digitally displayed in the passenger carriages. By contrast, in type 69 trains the drivers announce the stations through the speaker system.

Worker competencies
The workers' competence needed is to know the procedures for informing passengers so well that they are able to perform them while driving. The question of what to say is specified in the traffic safety manual and is a part of their formal education.

3) Maintain Safety of People and Equipment
Higher order strategy
Maintaining the safety of the train and the people along the tracks is the third higher order strategy and is associated with the subtask of approaching a station. It is also found in the other two subtasks.

Lower order strategies
The lower order strategies are to monitor the tracks and the environment for disturbances and to be prepared to brake. These lower order strategies are also found in the two other subtasks. Making sure that the station is clear of other trains is specific to the subtask of approaching a station. The following description only applies to this subtask, as the other two subtasks have already been described.

Social organisation and cooperation
The signal system is responsible for letting the driver know that the station is clear to enter. The ATC helps to ensure that the driver does not pass this signal until it is safe to do so, and the traffic control centre takes over responsibility for this in the case of malfunctions.

Worker competencies
The human capability required is attention. In addition, the drivers' procedural and environmental competencies are important. The procedural competencies, both formal and informal, are to know what to do in the case of malfunctions. The environmental competencies are more informal and are concerned with where the signals are, whether drivers' can expect a stop sign, etc.

4) Stop smoothly and safely
Higher order strategy
The fourth higher order strategy is to stop the train smoothly and safely. Most of the drivers interviewed informed that they invest quite a lot of effort in making the train travel as smoothly as possible to increase passenger comfort. Another technique for stopping the train safely is not to approach the station too fast.

Lower order strategies
A gradual decrease in speed is necessary when approaching a station. However, this decrease needs to be balanced against the demands of the timetable, because a premature decrease in speed can cause delays.

Social organisation and cooperation
The drivers are the most important actors in stopping the train smoothly and safely. The train controls the actual decrease in speed and the speedometer gives information about the changes in speed, but it is the drivers who execute the task.

Worker competencies
Worker competencies and especially experience are crucial in stopping a train smoothly and safely. The process is affected by many factors that need to be balanced. Knowing the effect of the brakes is one example, and this information is available for the drivers in the technical report, which follows every train. They will also know the braking length associated with the particular train type. The weather also affects the braking curve of the train. All of these factors are part of the calculations that drivers have to make in order to stop the train smoothly. This demonstrates the procedural, environmental and technical competencies needed. Many of the drivers interviewed also reported this competency as something that distinguishes more highly skilled drivers.

5) Maintain safety of people at the station
Higher order strategy
The next higher order strategy is to maintain the safety of the people on the station. The station area can have a lot of people near the tracks. Driving into a station and letting people on and off the train requires increased focus on the safety of these people.

Lower order strategies
The lower order strategy is to monitor the behaviour of people on the station.

Social organisation and cooperation
The social organisation and cooperation for this higher order strategy are divided between the driver, equipment and the conductor. By conductor, I refer to a conductor with a safety clearance who is responsible for the safety of the passengers on-board (hereafter referred to as conductors. Both train types (69 and 72) are required to have this type of conductor on-board. The driver and the conductor collaborate to maintain the safety of the passengers as well as the people on the station. This allocation of responsibilities and cooperation is described in the procedures. In addition, the driver and the conductor are required to brief each other on important information on every trip or work shift. The driver is responsible for the people on the station as long as the doors have not been opened. When the train is approaching a station, the driver monitors the station both forwards and to the rear (using the rear-view mirror). At some stations, drivers are required to sound the horn to signal the 'train is coming' before entering the station area. A sign by the tracks prompts drivers when to sound the horn. Sometimes the driver also warns people with the horn if they are close to the gap between the train and the platform. After the train has stopped and the doors are opened the conductor or guard has the main responsibility for the safety of people on the station, though the driver will still keep an eye on the situation in the rear-view mirror.

Worker competencies
Attention is crucial for maintaining safety. The drivers draw on procedural competencies from their training and social and environmental competencies from their experience. Some people such as children, or adolescents and/or passengers with a high alcohol intake, display more risk behaviour than others. Hence, drivers will be extra cautious at stations with schools in the vicinity, or at the weekends. The weather also plays a part and can, for instance, make the platform slippery.

6) Let people on and off
Higher order strategy
Letting people on and off the train is a higher order strategy realised through several lower order strategies.

Lower order strategies
When people are let on or off a train the doors need to be opened accompanied by an announcement and the process needs to be monitored.

Social organisation and cooperation
The driver opens the doors and announces that the doors are opening. In the type 72 train the automatic announcement system is responsible for the announcement. The process is monitored by the driver in the mirrors, and also by the conductor on the platform. It is the conductor who has the main responsibility for looking after the passengers when letting people on and off. On type 72 trains the driver can also check the IDU screen for information about the doors, but observations during the study have shown that they seldom look at the screen at this point.

Worker competencies
The driver needs to be attentive, and know the procedures relating to opening the doors and what his/her responsibility is. The driver also benefits from experience when it comes to environmental competencies, such as being aware of a slippery platform, and social competencies, such as understanding small children's behaviour.

7) Make Sure Equipment Works Adequately
Higher order strategy
The last higher order strategy associated with the subtask of approaching the station is to make sure that the train and equipment work adequately.

Lower order strategy
This is done by the lower order strategy of monitoring the equipment inside and outside the driver's cab, and this strategy can be found in all of the three subtasks. However, examples of equipment especially relevant for this particular subtask are the brakes, the doors, and the automatic announcement system. The latter can

announce the stations at the wrong time. In this case, the driver is required to detect this malfunction, turn off the system, and take over responsibility for making announcements. The following description only applies to the particular strategies for this subtask, as the others have already been described.

Table 2.3 Approaching station - the subtask of safely, comfortably, and efficiently stop at station to let people on/off.

Control task		Strategies		Sos. and Org	Worker competencies
Overall task	Subtask	Higher order strategies	Lower order Strategies		
Safely, efficiently, and comfortably transport passengers from a-b on time	Safely, comfortably, and efficiently stop at station to let people on/off	1) Detect station	Monitor the environment	Driver Route book Signal & Signs Automatic station announcer (72)	Human capabilities – Attention Competencies – Procedural – Environmental
		2) Inform costumers about station	Announce next station	Driver Automatic station announcer (72) Speaker (69) In cab text-sign (72)	Competencies – Procedural
		3) Maintain safety for people/equipment on tracks	Monitor out cab environment Make sure that the station can be entered Signal to others if danger	Driver Mirror Signals & Signs Horn ATC Train	Human capabilities – Attention Competencies – Procedural – Environmental
		4) Stop smoothly and safely at the right time	Adjust the speed by decreasing engine power and initiating brakes	Driver Signals & Signs Train Speedometer	Competencies – Technical – Procedural – Environmental
		5) Maintain safety for people on station	Monitor the behaviour of people on the station Monitor area passed the train Signal before approaching certain stations	Driver Mirror Horn Brakes Conductor Signs	Human capabilities – Attention Competencies – Procedural – Environmental – Social
		6) Let people on/off	Open doors, and announce that doors are open Monitor exit and entrance of passengers	Driver Automatic announcer Conductor IDU	Human capabilities – Attention Competencies – Procedural – Environmental
		7) Make sure train/equipment works adequately	Monitor train/equipment	Driver IDU Lamps Mirrors DROPS	Human capabilities – Attention Competencies – Technical – Procedural – Environmental

Social organisation and cooperation
The driver has the main responsibility for making sure that equipment works adequately. S/he has to check that the announcement system is announcing the stations correctly. The IDU screen assists in monitoring the technical system, and it also has a function that shows the status of the doors.

Worker competencies
The worker competencies are technical, procedural and environmental competencies, and have already been described in detail in the corresponding section under the heading '*3) Make Sure Equipment Works Adequately*'.

3.2.4 Leaving a Station

The subtask of safely, comfortably and efficiently leaving a station on time can be described in five higher order strategies: 1) Maintain safety of people on station, 2) Leave station on time, 3) Reach travel speed fast, 4) Maintain safety of people/equipment on tracks, 5) Make sure train/equipment works adequately. For an overview see Table 2.4.

1) Maintain safety of people on station
Higher order strategy
The higher order strategy is to maintain safety of the people on the station when the train is leaving the station.

Lower order strategies
The lower order strategy is to monitor the behaviour of the people on the station. It is important to watch areas both backwards and forwards.

Social organisation and cooperation
This is primarily the conductors' task when the doors are open, and the drivers only contribute by monitoring the mirrors. However, the drivers have sole responsibility as soon as the doors are closed.

Worker competencies
The main human capability relevant here is attention. Drivers also need to have environmental competencies such as knowledge about slippery platforms. Social competencies such as knowledge about people's behaviour can also be important.

2) Leave station on time
Higher order strategy
The second higher order strategy is leaving the station on time, and is important in order to keep to the schedule. Two aspects of this higher order strategy are especially critical: the train should only leave when it is safe to leave, and the passengers entering the train should be safe.

Lower order strategies
Leaving a station on time involves a sequence of five lower order strategies. First, it is necessary to find out whether it is safe to leave the station (i.e. that there are no other trains on the tracks ahead). Second, it is important to make sure that the passengers are safe. Third, the doors need to be closed. Fourth, the train is ideally required to leave the station exactly on schedule. Finally, the train must accelerate as fast, comfortably and as safely as possible.

Social organisation and cooperation
The allocation of the strategies described above are distributed between the driver, the conductor or guard, the signal system, the IDU screen, mirrors, the train, the route book, and the clock. The departure routine describes the organisation of these actors.

First, the information about when it is safe to leave the station is obtained through the signal system and it is up to the driver to collect this information. The driver communicates the 'permission granted' signal to the conductors via a lamp outside the train. The conductors double-check this information. Second, it is the conductors who decide when the train can leave, and they signal this to the driver with a flag or lamp. Their responsibility is to make sure that the passengers are safe, that the train has permission to leave and that it leaves on time. Third, the drivers can close the doors when the conductors have signalled that they are ready to leave. The drivers of type 69 trains watch the doors close in the mirror, while in type 72 trains the IDU screen serves as an addition aid for the drivers. The doors of type 72s are actually closed two times. This is because all of the doors are closed except the conductor's door. The conductor needs to make that everything is alright before going inside and close his door. The driver is next required to lock the doors. A message relayed by a speaker system informs the passengers that the doors are closing in order to ensure passenger safety. The automatic announcement system takes care of this on the type 72 trains. Fourth, the conductor checks the time before giving the signal to leave. The driver needs to check the clock against the route book to ensure that the train leaves on time. Drivers on the inner-local route (type 69 trains) reported that their tight schedule often makes checking the time redundant because it is all about doing things as fast as possible regardless. Finally, the driver has to start the acceleration and monitor the platform area.

Worker competencies
This sequence entails a large number of worker competencies. The most salient human capability here is attention. The competencies required are procedural, technical, environmental, and social. The formal technical knowledge is connected to getting the train to move and operating the other equipment. The procedural knowledge relates to the procedures, both formal and informal, for leaving a station safely. Environmental knowledge relates to the characteristics of the station, weather and traffic. The social knowledge is associated with the characteristics of the conductor or guard and the behaviour of the passengers boarding the train.

3) Reach travel speed fast

Higher order strategy
The third higher order strategy for leaving a station is to reach the travel speed fast. This is instrumental in keeping a tight schedule.

Lower order strategies
The lower order strategy is to accelerate as fast as possible, in a safe and comfortable manner.

Social organisation and cooperation
The signal system sets the maximum allowed speed, but the drivers are responsible for the actual execution of the lower order strategy of accelerating as fast as possible.

Worker competencies
The human capability that is most involved in reaching travel speed fast is attention, though the procedural competencies associated with driving the train are also important. This part of the task is a part of drivers' formal training in rules and procedures. Their informal knowledge involves knowing how to operate the train as efficiently and smoothly as possible. The environmental competencies include incorporating information about the weather, the train's technical status, and features of the line.

4) Maintain safety of people on the station

Higher order strategy
The fourth higher order strategy is to maintain the safety of people and equipment on the station while leaving the station.

Lower order strategies
The lower order strategy applied when leaving the station is monitoring the tracks and the environment for disturbances both forwards and backwards.

Social organisation and cooperation
The driver is the only one responsible for social organisation and cooperation.

Worker competencies
Monitoring the safety of people and equipment on the station requires attentiveness on the part of the driver. Social and environmental competencies are involved too, given that schoolchildren take more risks, and some passengers cross the tracks after getting off the train. The environmental competencies include, for example, knowing that a slippery station platform can lead to latecomers falling onto the tracks.

5) Make sure train/equipment works adequately

Higher order strategy
The final higher order strategy for leaving a station is making sure that the train and equipment function adequately.

Lower order strategy
The higher order strategy is carried out by the lower order strategy of monitoring the equipment inside and outside the driver's cab. This strategy can be found in all of the three subtasks. The equipment that is especially relevant for this particular subtask comprises the doors, the engine, and the automatic announcement system. The following description only applies to the typical strategies for this subtask, as

Table 2.4 Leaving station - the subtask of safely, comfortable, and efficiently leave the station on time.

Control task		Strategies		Sos. and Org	Worker competencies
Overall task	Subtask	Higher order strategies	Lower order Strategies		
Safely, efficiently, and comfortably transport passengers from a-b on time	Safely, comfortably, and efficiently stations on time	1) Maintain safety for people on station	Monitor platform	Driver Conductor	Human capabilities – Attention Competencies – Environmental – Social
		2) Leave station on time	Make sure that it is safe to leave Make sure passengers are safe Close doors Start accelerate comfortably, safe and efficient	Driver Signals Conductor IDU Route book Train	Human capabilities – Attention Competencies – Procedural – Environmental – Social
		3) Reach travel speed fast	Accelerate as fast as possible	Driver Signals & Signs Train	Human capabilities – Attention Competencies – Procedural – Environmental
		4) Maintain safety for people and equipment on station	Monitor the tracks and the environment for disturbances Monitor train/equipment	Driver Mirrors Conductor	Human capabilities – Attention Competencies – Procedural – Environmental – Social
		5) Make sure train/equipment is functioning adequately	Monitor train/equipment	Driver IDU Lamps Mirrors	Human capabilities – Attention Competencies – Technical – Environmental

the others have already been described in detail in the corresponding section under the heading '*3) Make Sure Equipment Works Adequately*'.

Social organisation and cooperation
On type 69 trains, the drivers use mirrors to see whether the doors have closed properly, whereas on type 72 trains, drivers are aided by an IDU screen, which shows whether the doors are open. The automatic announcement system gives information about the closing of the doors, and the drivers need to check that the information is correct.

Worker competencies
The driver needs to be attentive and make sure that everything is as it should be. The competencies required are technical and environmental. The technical competencies are informal as well as formal, and an example of such competencies is knowing how the fast the train is supposed to accelerate under given circumstances, such as slippery tracks.

4 Discussion
4.1 General Findings

The results outlines the many details of the train driving task. At the same time, several features were also revealed. The tasks have been presented in terms of the work domain, control tasks, strategies, social cooperation and organisation, and worker competencies. The overall task of train driving is structured into relatively clearly defined subtasks. Each of these subtasks shapes the strategies available, the social organisation and cooperation, and the demands made on the drivers. These general features will be discussed in the following section.

4.1.1 Task Transformation and Constraints

Train drivers' work conditions are characterised by rapid change. The conditions are altered as they constantly move through the landscape and switch between the different phases or subtasks of driving, and deal with the variability of the technical and organisational components. This feature is found in other transport systems (Petersen & Nielsen, 2001). The work domain sets the frame for the task in question. In the analysis this framing can be seen when the different control tasks, strategies, social cooperation and organisation, and worker competencies change as the work domain changes. To give an example, the weather, which varies significantly in Norway, influences the task of train driving in many ways. For instance, the ability to maintain the right speed (acceleration and deceleration) is subject to variation as snow, ice and rainfall affect friction. Less friction alters the train's braking capabilities, and thus alters safe speed limits. This is something that must be incorporated into the lower order strategies. It also changes the social organisation and cooperation, as the drivers rather than the signs decide the maximum speed. This in turn forces the driver to compensate for the mismatch between the fixed sign system and the variable weather.

The control tasks have been described in terms of one overall task and the three subtasks: 1) on the line, 2) approaching a station, and 3) leaving a station. This three-way split of the train driving task has been found in other studies too (Brotnov, 2007; Kecklund, 2001; Jansson et al. 2006; Jansson, Olsson & Kecklund, 2000). The different subtasks also shape the lower layers of the conceptual distinctions. For example, the higher order strategy of making sure the equipment is functioning adequately is important in all three subtasks. However, the driver monitors and pays attention to different parts of the system during each subtask. Thus each subtask can be regarded separately, both in serving to structure the work of the driver in practice, and for the purpose of analysis.

The different strategies shape the lower level of social organisation and cooperation by defining the control tasks involved and how they are to be allocated and coordinated. For example, monitoring speed and calculating the braking curve is a strategy divided between the ATC and the driver. Worker competencies are described last and workers deal with the constraints of all the five conceptual distinctions.

The results demonstrate how the general task can be broken down into concrete demands through the structure borrowed from CWA (Vicente, 1999). Similarly, Leplat (2000) argues that general tasks given to operators always are deconstructed down to concrete actions. This means that the emphasis is not on how to execute the work task, but how to adapt to or accommodate it. Leplat further argues that this task transformation is made in relation to how the workers understand the task. However, the workers' understanding of the task may not necessarily be correct. For example, drivers were found to have the wrong idea about how the ATC works (Jansson et al. 2000). Thus, dangerous situations or accidents can occur if the adaptations do not respect the constraints of the task (Christoffersen &Woods, 1999). Safety can then be thought of as respecting constraints (Vicente, 1999). This CWA-based paper has provided an overview of what the train driving task has to adapt to, namely the constraints. I have shown how the constraints affect the drivers' possibilities for taking action. In this sense, the formative analysis has served as a frame of reference for task transformation.

4.1.2 The Driver's Role: Dealing with context-conditioned variability

'It is not our problem. I don't mean that it isn't our problem ... It is not our fault, but it it's ours to fix.' (A driver's reaction to the many disturbances that have to be dealt with in the course their work)

Two points can be made about the drivers' role in the train driving task. First, the train driving task is heavily constrained in terms of the physical space in the cab, the safety equipment, the many rules and regulations, and the many procedures. Almost every subtask and circumstance that is experienced is covered in the train safety manual. The drivers themselves describe their task as highly regulated. However, Leplat (1994) argues that every prescription of a task implies that the worker must compensate for the shortcomings. This brings us to the second point. The analysis shows that the drivers on the local lines have to deal with a high degree of context-conditioned variability and they need to adjust to the different constraints and change their actions in order to achieve the set goals. The rationale for the presence of humans in work systems is both their ability to manipulate objects and tools in a flexible and versatile way as well as being adaptive problems solvers (Pejtersen, 1995). The timing and interpretation of the work situation precedes the actions chosen. The train safety manual does not formulate solutions for situation-specific variance, and to a large extent the driver's experience and expertise serve as a condition for the implementation of safety measures. An important value in the formative approach is the ability to identify and support workers' adaptive behaviour (Vicente, 1999). Three examples can be used to demonstrate how the drivers deal with the context-conditioned variability.

First, to stop a train smoothly is dependent on and constrained by several different factors. The driver applies the braking system to bring the train come to a halt smoothly. The application of the braking system is used in coordination with a number of information resources: signs informing the distance to the next station, topographical variations along the tracks, weather status, technical status of the train, and so forth. None of these information sources or conditions is sufficient in themselves to guide an approach to a specific station, but they are used in combination to accommodate safety, efficiency and comfort.

Second, a significant source of the context-conditioned variability experienced by drivers is failure in the technical equipment. Both type 69 and 72 trains experience regular technical problems, and a failure on the heavily trafficked lines can lead to delays for up to a whole day. If a problem occurs the drivers are required to try to locate and fix it. One such example is when the automatic mode of the type 69 train does not work after the train turns at an end station. This is an easily solved and quite common error, which is caused by a loose fuse. The drivers then have to find the right fuse cabinet and push the fuse back in. The drivers take great pride in successfully diagnosing a problem and/or repairing a fault. In England, train drivers are known as 'train engineer' or 'locomotive engineer', and these titles emphasise the technical expertise required for their work. The aforementioned example shows how the technical equipment is both a condition for solving the task and the focus of problem solving. Thus, the technology is not objective and stable, but in itself is a source of variability. In other words, the drivers are required to drive as well as make sure that their vehicle works.

Third, the signal system is responsible for controlling the traffic on the line and ensuring safe operations on a busy line. Trains are not allowed to pass a signal that displays the message 'stop'. A malfunctioning signal is programmed to stay in 'stop' mode until the problem is resolved. While this prevents the dangerous situations of 'signals passed at danger' (SPADs), it can also halt the traffic flow and cause delays. Drivers who come across signal which have been 'in stop' for too long are required to call up the traffic control centre, get permission to drive past the signal, reset the ATC, and continue their journey. This is described in detail in the traffic safety manual, but the actual evaluation of what is 'too long' is dependant on the drivers' knowledge about the route and the traffic on the day in question. These examples show

how the drivers deal with the context-conditioned variability and how the system is dependent on the drivers' adaptive actions in order to achieve its goal. They also point to the technical system as a source of variability in itself, which influences the allocation of tasks. This means that social organisation and cooperation involved in the task is not only altered by strategies, control tasks and work domain but also technology.

In contrast to the normative approach, the formative approach is explicitly aimed at work characterised by context-conditioned variability. The findings show that the task has a high degree of context-conditioned variability and this thus supports our choice of analysis.

4.1.3 Route Knowledge

Train drivers deal with context-conditioned variability when they drive a train. Their competencies enable them to handle these challenges. For the analysis in this study, worker competencies were divided into human capabilities and competencies (environmental, technical, procedural, and social) (Vicente, 1999). Another distinction that was made was between formal and informal competencies. While formal competencies are a prerequisite for driving the train safely, extensive informal competencies acquired through experience are also important. For example, choosing the right speed is a higher order strategy that is dependent on the drivers' capacity to incorporate several factors. The formal competencies required here are associated with knowing the rules and regulations, and the technical aspects of operating the train. There are many examples of the informal competencies needed; for example, knowledge about stations with schoolchildren or where tractors are likely to be working near the tracks is important for safety. Economic driving is dependant on, for example, knowing in advance that the tracks go downhill in a few minutes, which makes an increase in speed unnecessary. Passenger comfort is also increased when the driver knows that one should start to decrease speed before passing e.g. a specific topographical feature.

Informal competence has also been investigated under the term 'route knowledge', and is defined by being experience-based, gained from many types of information sources and necessary to operate safely and efficiently (Biemans, 2006). Swedish studies of the train driving task have also looked at route knowledge, describing it as a critical part of the drivers' operational knowledge base (Jansson et al. 2006; Jansson et al. 2000). The use of informal knowledge in the present study is more detailed than the broadly defined route knowledge. I specify whether the knowledge or competence is linked to the performance of a task (procedural), topography, features of the track, or weather (environmental), technical issues (technical), or to people's behaviour (social).

The importance of drivers' route knowledge or informal competencies is supported by the findings. It also shows how the ability to deal with the demands of context-conditioned variability successfully is closely linked to experience-based competencies.

4.2 Evaluation of the Formative Approach

In this study, the application of the formative approach on the train driving task was useful as it provided an overview of the task. The findings suggested how the constraints associated with the higher conceptual distinctions shaped the lower ones, and ultimately the task as a whole. The findings also revealed how the train driving task is characterised by context-conditioned variability. In addition, the importance of informal competencies was emphasised. Some difficulties with applying the approach were encountered, however, and these will be discussed in the following.

4.2.1 Device-dependence

A clearly stated goal of formative analysis is to avoid the task-artefact cycle. This is solved in the CWA framework by describing the task without referring to the existing technical solutions aiding the work. In other words, the analysis aims to avoid device-dependence (Vicente, 1999). The definitions of the conceptual distinctions clearly state that the work domain, control tasks, and strategies should be referred to without mentioning any technical solution. It is in the description of the social organisation and cooperation that these features of the system relate to.

Nevertheless, it proved difficult to describe the work domain, control tasks, and strategies without referring to the technical solutions. This could be due to the fact that train driving is inherently device dependant (i.e. upon the train), and that the very premise of describing parts of the task without referring to the device therefore is difficult. Train driving is a task aided by automation, and this posed a problem in the analysis. This is evident from how the analysis is characterised by the occasional jump in the level of analysis, and one example is how the drivers maintain the right speed aided by the ATC. This solution affects the drivers' lower order strategy of keeping the right speed, and to have excluded this in the presentation of these strategies would have made the analysis incomplete. The description of the ATC conceptually belongs at the next level, namely the social organisation and cooperation, and in this manner is not consistent with the definitions in the analysis.

Lind (2003) discusses the problems associated with analysing a system independent of automation. He argues that CWA (Vicente, 1999) definition and modelling of the conceptual distinctions as independent of the automated systems makes an artificial distinction between systems with and without automated technical solutions. The implication of this is that the modelling of conceptual distinctions cannot be used in interface design in systems with automated control devices. To have excluded these devices from the analysis would have decreases the value of the study in terms of design implications, which is a clearly stated goal of the formative analysis. Lind (2003) further states that it is important that Cognitive Engineering meets the needs of the real world, and that a design methodology should be applicable even though total device-independence is not possible. Burns and Hajdukiewicz (2004) also found device-independence difficult to achieve in their analysis of the work domain. Their article further states that it can be important in such analyses to include the state of the automated system in relation to other components. They suggest dual models,

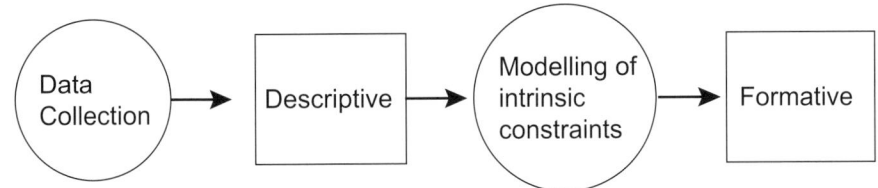

Figure 2.2 The relation between the descriptive and the formative approaches. Adapted from Vicente (1999) p. 133.

whereby the automated system is taken into consideration on one model, and not in the other.

I have argued earlier that the main weakness of the descriptive approach is the device dependence. The description of the task presented in this paper is device-dependent. This is a serious flaw because one of the arguments for making a formative analysis is the normative and descriptive approaches' inability to escape the task-artefact cycle. It can thus be argued that this paper is descriptive rather than formative. However, even Vicente (1999) accepts to a certain degree that the description of a task is descriptive, and what makes an approach formative is the modelling of the behaviour shaping constraints (see Figure 2.2). This paper has described the behaviour shaping constraints, and provided an overview of the contraints of the train driving task usable for further modelling.

An analysis will be less efficient if it is not able to free it self from the existing solutions, because it will inherit whatever deficiencies are present in existing design (Vicente, 1999). This means that the analysis presented in this paper will inherit the deficiencies of the present train driving task. However, Burns & Hajdukiewicz (2004) argue that a work domain model is, in a way, a recreation of some of the steps behind the original design of the work domain. This idea can also apply to the formative approach as a whole. The formative analysis can, even if the goal of total device-independence is not attained, clarify some of the steps behind the original design. This in turn can help to clarify the interdependence between the different components of the system and prove informative when altering existing technology, and adding new or changing other parts of the system.

Device-independence was not achieved in this study. The value of describing a task as device independent is an unresolved issue and should be discussed further. In the end, it is the modelling of the behaviour shaping constraints that make an analysis formative. This paper fills this demand, and the analysis can provide a recreation of some of the steps behind the present work solution. In this respect, the analysis may be device-dependent and escapes some of the existing deficiencies.

4.2.2 Work as a Continuous Flow of Action

The formative analysis applied here shows the train driving task as a string of separate subtasks, higher and lower strategies, social organisation and cooperation, and worker competencies. In reality, the different subtasks and their subsequent paths are a part of a continuous flow of action. One example is how train drivers initiate the departure routine by checking for the signal for permission to leave while approaching the station. Another example is how they often give the conductor or guard the 'permission granted' signal and open the doors simultaneously. This flow of action was illustrated by the operational sequence diagrams (OSDs) (see Appendix 1 for examples).

The disintegrated presentation is a result of how the formative analysis models the task. Studies of both Cognitive Systems Engineering (e.g. Hollnagel, 2001; Hollnagel & Woods, 2005; Woods & Hollnagel, 2006) and Action Theory (Nardi, 1997; Norros, 2004) have discussed the pitfalls of a disintegrated view of work in action. What makes a formative analysis formative is the description of the behaviour shaping constraints, and fragmentation of the task is necessary in order to show these. Despite the disintegrated presentation of the flow of action, the results are made clear by grounding the paper in observations of work in context, and using techniques such the OSD.

4.2.3 Level of detail in the analysis

This paper has been based on CWA but has not made a complete analysis. The modelling tools and the design interventions have been cut out from the application of the formative approach. From the beginning of the study it was decided to sacrifice the level of detail for the broader picture of the relationship between the conceptual distinctions, and in order to be explorative. However, deciding where to set the actual level of detail when analysing the task proved difficult, as will be shown by the three examples discussed in the following.

First, this analysis presents some of the strategies present in the train driving task. They do not represent an exhaustive list of all of the available strategies because several strategies may be available (Vicente, 1999), and the scope of this paper did not allow for such detail. The ones presented here are largely those found in the existing organisation of work, and the results discussed earlier may be described as descriptive. Second, the study was first and foremost conducted from the drivers' point of view, and this is important to take into consideration when evaluating the approach. This narrow perspective becomes especially evident in the analysis of the level of social organisation and cooperation. The allocation of work has been presented here, but the collaboration between the actors has not been treated in detail. Third, to decide on the level of detail of worker competencies also proved difficult. This level deals with human capabilities and limitations, and hence is a category that deals with the whole field of psychology. However, attention has been mentioned as it is important in the monitoring strategy.

The problem of deciding on the level of detail was also linked to the choice of not applying any of the modelling tools suggested by CWA. Despite these problems, this paper has provided a description of how the constraints on the different conceptual distinctions influence each other and shape the task. This was what was hoped to be gained from applying this broad formative analysis. Following from the explorative nature of this paper, it is also possible to identify weaknesses in the analysis of information about interesting areas for further investigation.

4.3 Further studies

This paper has outlined how constraints in conceptual distinctions shape the train driving task. The CWA framework (Vicente, 1999) suggests the use of modelling tools associated with these conceptual distinctions, and these provided more detail. In future studies, the findings relating to the use of these modelling tools could be incorporated in the framework suggested by this paper, in order to further investigate how constraints shape the train driving task.

One of the aims of the formative approach is to support workers dealing with context-conditioned variability. The results show that this is an important feature of the train driving task. Devices such as the IDU screen, DROPS centre and Traffic Control Centre, and also the conductors are important forms of support for drivers when confronted with the context-conditioned variability. An interesting focus for further studies could be on how the organisation and devices aid train drivers in dealing with context-conditioned variability.

4.4 Conclusions

This paper applied a formative analysis of the train driving task. It has shown how the different constraints in five conceptual distinctions shape the train driving task. The analysis also sheds light on how train drivers deal with context-conditioned variability. The formative approach applied here thus proved itself to be useful for the analysis of train driving. Some problems were encountered. These were concerned with the ability to achieve complete device-independence and portraying the work as a continuous flow of action, and there were difficulties in deciding on the level of detail. Nevertheless, the paper provides an alternative starting point for formative analysis by drawing a broader picture emphasising the interactions of the conceptual layer, and thereby it contributes to Cognitive Engineering by offering a coherent framework for the presentation of formative analysis.

Table 2.5 Explanations of icons used in the Operational Sequence Diagrams (OSD) of approaching and leaving station.

Operational Sequence Diagrams (OSD) of approaching and leaving station			
▽	= Information given	▼	= Information assumed given
☐	= Action / control action	●	= Information assumed received
○	= Information received	✿	= Assumed previously stored information
○	= Automatised information		*Assumed* = attributed on the basis of interviews, direct observations, and cognitive walkthroughs

References

Biemans, M. (2006). Cognition in Context: The effect of information and communication support on task performance of distributed professionals. Enschede: Telematica Instituut.

Brotnov, S.M. (2007). Railway Driving Operations and Cognitive Ergonomics Issues in the Norwegian Railway: A Systems Analysis. Unpublished Master's Thesis, University of Oslo, Oslo, Norway.

Burns, C. M. (2004). Lessons from a Comparison of Work Domain Models: Representational Choices and Their Implications. Human Factors, 46 (pp. 711-727).

Burns, C. M. & Hajdukiewicz, J. R. (2004). Ecological Interface Design. London: CRC Press.

Bødker, S. (1991). Through the Interface: A human activity approach to user interface design. Hillsdale, US: Lawrence Erlbaum Associates.

Carroll, J.M., Kellogg, W.A. & Rosson, M.B. (1991). The Task-Artifact Cycle. In Carroll, J.M. (ed). Designing Interaction: Psychology at the human-computer interface. (pp. 74-102) Cambridge, UK: Cambridge University Press.

Casey, S.M. (1998) Set Phasers on Stun: And other true tales of design, technology, and human error. Santa Barabara, US: Agean.

Christoffersen, K. & Woods, D. D. (1999). How Complex Human Maschine Systems Fail: Putting "Human Error" in Context. In Karwowski, W. & Marras, W.S. (eds). The Occupational Ergonomics Handbook. London: CRC Press.

Dekker, S.W.A. & Woods, D.D. (2002). MABA-MABA or Abracadabra? Progress on Human- Automation Co-ordination. Cognition, Technology & Work. 4 (pp. 240-244).

Eurocontrol (2003). Core Requirements for ATM Working Positions. Capture and Exploitation of Requirements. HRS/HSP-006

FOR 2002-01-29 nr. 122: Forskrift om offentlige undersøkelser av jernbaneulykker og alvorlige jernbanehendelser. Accessed November 24, 2005 from http://lovdata.no/for/sf/sd/xd-20020129-0122.html.

FOR 2002-01-29 nr. 123: Forskrift om varslings- og rapporteringsplikt i forbindelse med jernbaneulykker og jernbane hendelser. Accessed November 24, 2005 from http://lovdata.no/for/sf/sd/xd-20020129-0123.html.

FOR 2002-12-18 nr. 1678: Forskrift om krav til helse for personell med arbeidsoppgaver av betydning for trafikksikkerheten ved jernbane, herunder sporvei, tunnelbane og forstadsbane m.m. (helsekravsforskriften). Accessed November 24, 2005 from http://lovdata.no/for/sf/sd/xd-20021218-1678.html.

FOR 2002-12-18 nr. 1679: Forskrift om opplæring av personell med arbeidsoppgaver av betydning for trafikksikkerheten ved jernbane, herunder sporvei, tunnelbane og forstadsbane m.m.(opplæringsforskriften). Accessed November 24, 2005 from http://lovdata.no/for/sf/sd/xd-20021218-1679.html.

FOR 2001-12-04 nr. 1335: Forskrift om trafikkstyring og togframføring på statens jernbanenett og tilknyttede private spor (togframføringsforskriften). Accessed May 6, 2007 from http://sjt.no/www/Lover_og_forskrifter/filestore/Togframfringsforskriften/Togframforingsforskrift.pdf

FOR 2001-12-04 nr. 1336: Forskrift om signaler og skilt på statens jernbanenett og tilknyttede private spor (signalforskriften). Retrieved May 6, 2007 from http://sjt.no/www/Lover_og_forskrifter/filestore/Togframfringsforskriften/Togframforingsforskrift.pdf

Gautherau, V. & Hollnagel, E. (2005). Planning, Control, and Adaptation: A case article. European Management Journal, 23 (pp. 118-131).

Hoc, J. (2001). Towards Ecological Validity of Research in Cognitive Ergonomics. Theoretical Issues in Ergonomics Science, 2 (pp. 278-288).

Hollnagel, E. (2004). Barriers and Accident Prevention. Aldershot, UK: Ashgate.

Hollnagel, E. (2001). Time and Control in Joint Human-Machine Systems. People in Control: An international Conference on Human Interfaces in Control Rooms, Cockpits and Command Centres. Conference Publications No. 481.

Hollnagel, E. & Woods, D.D. (2005). Joint Cognitive Systems. Foundations of Cognitive Systems Engineering. London, UK: Taylor and Francis.

Hutchins, E. (1995). Cognition in the wild. Cambridge, MA: MIT Press.

Jansson, A., Olsson, E. & Erlandsson, M. (2006). Bridging the gap between analysis and design: Improving existing driver interfaces with tools from the framework of cognitive work analysis. Cognition Technology and Work, 8 (pp. 41-49).

Jansson, A., Olsson, E. & Kecklund (2000). Att köre tåg. Lokförarens arbete ur et systemperspektiv. Technical report 2000-031. Institutionen för informationsteknikologi, Uppsala Universituet, Uppsala.

Kecklund, L. (2001). Final report on the TRAIN-project. Risk and proposals for safety enhancing measures in the train driving system. Banverket, Borlänge.

Kirwan, B. & Ainsworth, L.K. (1992). A Guide to Task Analysis. Taylor and Francis, London.

Klein, G.A. (1995). A Recogniton-Primed Decision (RDP) Model of Rapid Decision Making. In Klein, G.A., Orsanau, J., Calderwood, R. & Zsambok, C.E. Decision Making in Action. Models and Methods. Norwood, US: Ablex Publishing

Leplat, J. (2000). Activity. In Karowski (ed). International Encyclopaedia of Ergonomics and Human Factors.

Leplat, J. (1994). Collective Activity in Work: Some Lines of Research. Le Travail Humain, 3.

Leplat, J. (1989). Error Analysis, Instrument and Object of Task Analysis. Ergonomics, 32.

Lind, M. (2003). Making sense of the abstraction hierarchy in the power plant domain. Cognition Technology and Work, 5. (pp. 67-81).

Maramas, N. & Nathanael, D. (2005). Cognitive Engineering Practice: Melting Theory into Reality. Theoretical Issues in Ergonomic Science, 6 (pp. 109-127).

Nardi, B.A. (1997). Articleing Context: A Comparison of Activity Theory, Situated Action Models, and Distributed Cognition. In B.A, Nardi (ed). Context and Consciousness: Activity Theory and Human-Computer Interaction. (pp. 69-102). Cambridge, Mass.: MIT Press

Norros, L. (2004). Acting Under Uncertainty. The Core-Task Analysis in Ecological Article of Work. Helsinki: VTT.

Norwegian National Railway Administration (n.d.). Norwegian National Railway Administration Homepage. Retrieved March 22, 2007 from http://www.jernbaneverket.no/

Norwegian National Railway Administration (2006). Norwegian National Railway Administration – Accident Statistics (2005). Retrieved May 5, 2007 from http://sjt.no/www/Publikasjoner/Ulykkesstatistikk/filestore/Publikasjoner/Statisitikk/rsstatistikk_ulykker/Ulykkesstatistikk_2005.pdf

NSB AS, Persontog Drift, (2007). Trafikksikkerhetsbestemmelser for togpersonalet i NSB AS (TS). (Traffic Safety Regulations for Train Personnel in NSB AS- Traffic Safety Manual).

NSB (n.d.). Norwegian State Railways Annual Reports. Retrieved March 22, 2007 from http://arsrapporter.nsb.no/NSB_Aarsrapport/CCS

Norwegian Railway Inspectorate (n.d.). Norwegian Railway Inspectorate Homepage. Retrieved March 22, 2007 from http://sjt.no/

Pejtersen, A.M. (1995). Cognitive Engineering in Information Retrieval Domains – Merging Paradigms? Bibliothek 19, 1.

Perrow, C.(1999). Normal Accidents. Living with High-Risk Technologies. Princeton University Press. Princeton, New Jersey.

Petersen, J. & Nielsen, M. (2001). Analyzing Maritime Work Domains. In: Onken (R (Ed.). Proceedings from the 8th Conference on Cognitive Sciences Approaches to Process Control, 24-26 September, 2001, Universität der Bundeswehr, Germany, Nubiberg (pp. 221-230).

Rasmussen, J. (1997). Merging Paradigms: Decision Making, Management, and Cognitive Control. In Flin, R., Salas, E., Strub, M.E., & L. Marting (Eds.), Decision Making under Stress: Emerging Paradigms and Applications (pp.67-85). Aldershot, England: Ashgate.

Rasmussen, J. (1990). The Role of Error in Organizing Behaviour. Ergonomics, 33 (pp. 1185-1199).

Rasmussen, J., Pejtersen, A.M. & Scmidt, K. (1990). Taxonomy for Cognitive Work Analysis. Risø- M-2871.

Rizzo, Pozzi, Save & Sujan (2005). Designing Complex Socio-technical systems: A Heuristics Schema based on Cultural- Historical Psychology. Proceedings of the 2005 Annual Conference on European Association of ergonomics, Greece, 71-89.

Sarter, N.B. & Woods, D.D. (2000).Learning from Automatisation Surprises and Going Sour Accidents. In Sarter, N.B. & Amalberti, R. (Eds.) Cognitive Engineering in the Aviation Domain 2000.(pp. 327-352). Mawah, US: Lawrence Erlbaum Associates.

Shepherd, A. & Marshall, E. (2005). Timeliness and task specification in designing for human factors in railway operations. Applied Ergonomics, 36. (pp. 719-727).

Statistics Norway (2006). Traffic Accidents with Persons Injured in 2005. Retrieved May 5, 2007 from http://ssb.no/emner/10/12/20/vtu/

Suchman, L.A. (1987). Plans and Situated Actions. The Problem of Human Machine Communication. Cambridge, Cambridge University Press.

Vicente, K.J. (1998). Commentary. An Evolutionary Perspective on the Growth of cognitive Engineering: The Risø Genotype. Ergonomics, 42. (pp. 156-159).

Vicente, K.J. (1999). Cognitive Work Analysis. Toward Safe, Productive, and Healthy Computer- Based Work. London, UK: Lawrence Erlbaum Associates.

Wilson, J.R. & Norris, B.J. (2006). Human factors in support of a successful railway: a review. Cognition Technology and Work, 8 (pp. 4-14).

Wilson, J.R. & Norris, B.J. (2005). Rail Human Factors: Past present and future. Applied Ergonomics, 36. (pp.649-660).

Wilson, J.R, Cordiner, L., Nichols, S., Norton, L., Bristol, N., Clarke, T., & Roberts, S. (2001). On the right track: Systematic implementation of ergonomics in railway network control. Cognition Technology and Work, 3. (pp. 238-252).

Woods, D.D. (1988). Coping with Compelxity: The Psychology of Human Behaviour in Complex Systems. In Goodestein, L.P (Ed.) In Task, Errors, and Mental Models (pp. 128-148). Taylor & Francis

Woods, D.D. & Hollnagel, E. (2006). Joint Cognitive Systems. Patterns in Cognitive Systems Engineering. London, UK: Taylor and Francis.

Woods, D.D. (1998). Commentary Designs are hypotheses about how artefacts shape cognition and collaboration. Ergonomics, 41. (pp. 168-173).

Zsambok, C.E. & Klein, G (eds) (1997). Naturalistic Decision Making. Mawah, US: Lawrence Erlbaum Associates.

Appendix

Operational Sequence Diagrams (OSD) of Approaching and Leaving Station

Kolbotn Station - Train type 69

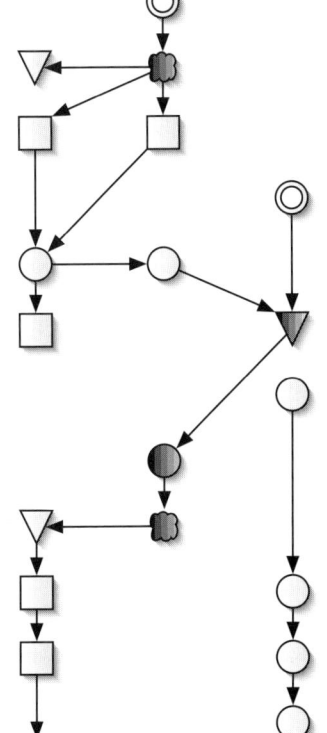

	Light signal - "clear"
Announce next station and platform side	Expecting next station (e.g. landmark - church)
Initiating brakes	Decreases speed
	Light signal - "clear"
Station area stop marks	Looks at station / mirror
Train at full stop	Signal "clear" received
	Looks at station/mirror
	Ready signal from conductor
Announce closing of the doors	Adapts departure and speed to timeschedule (late/on time/early)
Close doors	Light signal - "clear"
Release brakes and increase speed	Looks at station/mirror
	Looks forward

Out on the line

Embodied Minds – Technical Environments

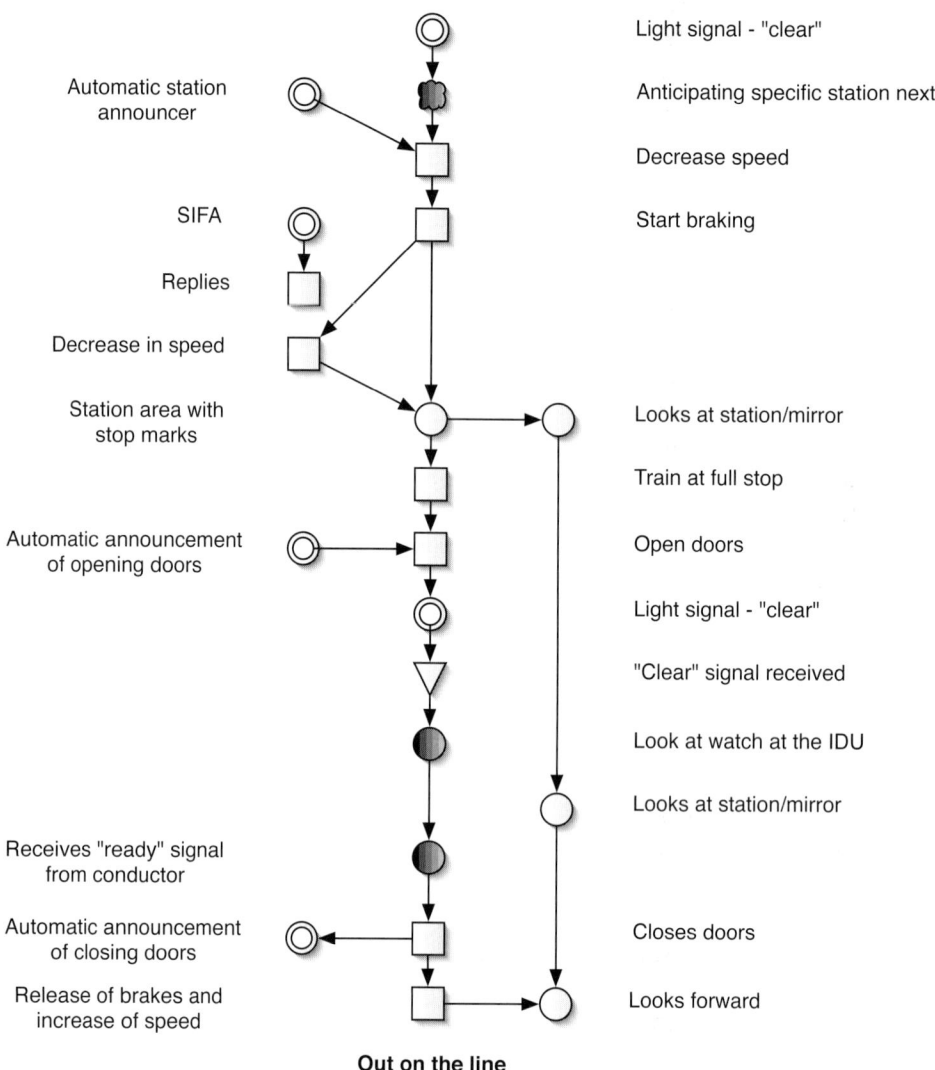

112

3

Railway Driving Operations and Cognitive Ergonomics Issues in the Norwegian Railway: A Systems Analysis

Sarah M. Brotnov

In this study, an initial systems-based analysis was performed within the Norwegian railway system. This approach to system safety attempts to provide information regarding the latent, system-related factors leading up to an unwanted event. The primary data include two semi-structured group interviews with safety officials, two semi-structured individual interviews with safety officials, one cognitive interview (CI) with a certified train driver, and two ethnographic observations with in-cab interviews. In addition, a total of 542 reports of Signal Passed at Danger (SPAD) were read and analysed. Leaving a station, out on the route, and approaching the station were found to be critical to train driving operations, while switching was suggested as a fourth system critical situation. The results suggest that there are several weaknesses within the Norwegian railway system which are likely contribute to unwanted events.

1 Introduction

Fatal crashes in the transport sector have a major impact on society, the domain at hand, and the involved stakeholders. The railway is not exempt from the ramifications of fatal accidents, but cognitive ergonomics/human factors[1] research in the rail domain has been somewhat in a state of neglect. Currently, a renaissance in railway human factors is taking place. Governments call for zero-tolerance policies, and while enterprises have managed to stabilise accident rates, there is still no evidence of a downward trend (Lawton, & Ward, 2005). Enterprises have expressed the wish to decrease potentially dangerous incidents and accidents, but actually reducing the rates appears to be challenging.

Modern risk analysis methods are based upon accident models that attempt to move away from human error as a central concept. As seen in CREAM (Cognitive Reliability and Error Analysis Methodology; Hollnagel, 1998) and a number of other human factors research publications, human error is regarded as a misleading, insufficient and incorrect concept (for more detailed discussion of the concept human error see, for example, Dekker, 2002; Hollnagel, 1983; Lawton & Ward, 2005; Rasmussen, 1990; Rasmussen, Nixon, & Warner, 1990; Reason, 1990; Vicente, 2004; Whittingham, 2004). Within this line of research, human performance variation is viewed as a natural, unavoidable, and positive part of any work task that is difficult to predict. It is rather when these valuable adjustments to work go awry that problems arise.

The railroad industry qualifies to be categorised as a complex socio-technical system (Wilson & Norris, 2006; 2005). A system is considered a complex socio-technical system if rated highly on several of the following dimensions: large problem space, social, heterogeneous perspectives, distributed, dynamic, potentially high hazards, many coupled subsystems, automated, uncertain data, mediated action via computers, and disturbance management (Vicente, 1999). It may even be argued that the railway system is perhaps the most complex industry existing (Shepard & Marshall, 2005). The railway industry stretches over a wide and ever changing geographical area and involves a wide range of engineering disciplines. In addition, customer satisfaction and safety concerns are necessarily taken into account. Different actors are located in different places, qualifying the system to be considered as 'distributed' (Wilson, Cordiner, Nichols, Norton, Bristol, Clark & Roberts, 2001): the system has drivers out on route, traffic controllers in the control room, company administrators at respective headquarters, education/training facilities, etc.

A systems analysis based upon models for understanding human behaviour in complex socio-technical systems (mentioned above) considers the safety of any system as a result of dynamic interactions at various levels of the Man-Technology-Organisation (MTO) triad. System models of man-machine interaction view the

---

[1] The terms cognitive ergonomics and human factors are used interchangeably throughout this paper and represent the more performance-oriented research tradition within ergonomics (Hollnagel, 1997).

persons and technical system as a *functional unit* and that work together to sustain control (Hollnagel, 2002; Hollnagel & Woods, 2005). Here, control is referred to as a product of the combination of Man and Machine and cannot be considered the product of only one of these in isolation. Control characterises the way in which one applies or uses competencies. To keep control refers to the ability the agent has to handle a dynamic process by reducing the effect of adverse conditions or situations which cannot be foreseen (Hollnagel 2002). Quite often the ability to have control is related to having sufficient time available to act to reduce adverse conditions or situations. It is important to mention that system models do not discern between man and machine, and the term 'cognitive' in this context refers to 'the ability to adapt to disruptions in the systems environment' rather than relating to thinking per se (Hollnagel & Woods, 2005).

Previous systems-based analyses have been shown to provide insight into vulnerable parts of the rail system, resulting in concrete safety recommendations for the railway (Lawton & Ward, 2005). Instead of finding human error, the goal of a systems-analysis is rather to provide insight as to how the system or situation contributes to conditions leading up to an unwanted event and to give explanations for why, for example, barriers failed to improve system safety (Dekker, 2002; Lawton & Ward, 2005; Rasmussen, 1990). Thus, by understanding the latent system-related factors leading up to an incident or accident, researchers and practitioners can either improve existing safety barriers or design new barriers in the system (Hollnagel, 2004). Latent system-related factors have previously been revealed when adapting system-based analysis to interpret factors that contributed to incidents and accidents where human-related causes, for example inattention, were previously found as a cause (Smiley, 1990; Van der Flier & Schoonan, 1988). There is, therefore, reason to believe that a systems analysis of the Norwegian railway can provide valuable information which can lead to improved system safety.

To be able to obtain valuable information for total system safety, it is necessary to understand the work task (Shepard & Marshall, 2005; Woods & Cook, 2002). The way in which drivers use information while performing work tasks is of the essence when designing safe systems. A thorough understanding of the work task is important when considering all human factors related issues, for example, when designing the in-cab interface, signal systems, and teaching aims and methods. In addition, task analysis can be referred to when considering issues of a more operational or managerial nature.

The Cognitive Work Analysis (CWA) framework is an approach to task analysis which demonstrates dimensions of the work task that need to be analysed if one wishes to design, implement or maintain the safety of a complex socio-technical system (Rasmussen, 1986; Rasmussen & Vicente, 1989; Rasmussen, Pejtersen & Goodstein, 1994; Vicente 1999). What is most important in this context is to understand the behaviour shaping constraints, or performance shaping factors, of the environment *in the actual work setting* (Hollnagel, 1998).

2 Scope and Purpose

The main aim of the project discussed in this paper is to provide an initial systems analysis of Norwegian State Railways (NSB (Norges statsbaner)) based upon modern cognitive ergonomics research. A variety of techniques will be used to assess the MTO relationship with respect to the train driving task in the Norwegian railway system. In this initial investigation, the perspective of the driver will be emphasised.

It is important to note that this project does not aim to present a complete (all-inclusive) systems analysis of the Norwegian rail domain. Rather, this investigation represents an initial attempt to look at the Norwegian railway service from a systems perspective. This report will have relevance for both practitioners and researchers. Although this is a research-based paper, it is important to keep in mind that the practitioner–researcher distinction is potentially risky when considering system safety (Hollnagel, 1998). Researchers need to use available knowledge of the task and how the system works in practice to be able to develop relevant models for behaviour and methodologies. At the same time, practitioners need relevant methods based upon psychological theory to be able to work with safety issues in the best possible way (Hollnagel, 1998; Woods & Cook, 2002).

This paper will 1) provide an understanding of the important system critical situations drivers are faced with, and 2) pinpoint some weaknesses observed in the Norwegian rail system. This initial study should provide a solid basis for the evaluation of effective resource allocation in the future.

In the first section of this paper I will attempt to find system critical driving situations in the Norwegian train driving task. How drivers use information will be discussed. Previously, the use of CWA in the Swedish rail industry has resulted in the train driving task being divided into three system critical driving situations (Jansson, Olsson & Erlandsson, 2006): *leaving a station, out on the route* and *approaching a station.*

The second section of this paper will discuss several rail human factors issues in light of the train driving task. The aim is not a complete evaluation of all human factors issues at all levels of organisation in the Norwegian State Railways, but rather an evaluation with the aim of pinpointing weaknesses in the system which may be ripe for improvement or further investigation. Some weaknesses that are pinpointed in this evaluation will need to be studied further before safe solutions can be found, while others may be strengthened with rather simple, cheap solutions.

3 Methods

In this study, qualitative methods were used, namely interviews, ethnographic observations with video material, critical incident report analysis, and analysis of secondary documents (Woods & Cook, 2002) and CWA (Vicente, 1999). The data were examined in the light of relevant human factors research and from the perspective of the train driver.

The primary data included two semi-structured group interviews with safety officials from Norwegian State Railways, two semi-structured individual interviews with safety officials from Norwegian State Railways, one cognitive interview (CI) with a certified Norwegian State Railway train driver, and two ethnographic observations with in-cab interviews with Norwegian State Railway personnel. The data were gathered within between September 2005 and September 2006. Several of the safety officials are also certified train drivers. The data included DVD material relating to the ethnographic observations with in-cab interviews and an audio recording of the CI. In the first ethnographic observation, a hand-held camera was used to record the in-cab environment, Man-Machine-Interaction, and in-cab interviews. In the first ethnographic observation the hand-held camera was also used to observe the outside environment, while in the second ethnographic observation a second stationary camera was used to record the outside environment.

Norwegian State Railways incident reports on Signal Passed at Danger (SPAD) events from 2003, 2004, and January–May 2005 were read and analysed. There were 251 incidents from 2003, 192 incidents from 2004, and 99 events from 2005, making a total of 542 analysed incidents. In addition, secondary data sources were used, such as official laws and regulations, various reports from the Norwegian Accident Investigation Board, Norwegian Railway Inspectorate, Norwegian Railway Authorities, American Federal Railway Association (FRA), and other relevant sources).

The collected data thus provide a background or basis for a research-based discussion about railway system safety issues. For the most part, the collected information will be referred to indirectly, to preserve informants' anonymity.

4 Results
4.1 Cognitive Work Analysis: Train Driving on Local Routes

Table 3. 1 lists examples of the three system critical driving situations and observable and non-observable actions. The results of the current CWA in the Norwegian railway system are in partial accordance with these findings. The three situations and their respective observable and non-observable actions were also found in the Norwegian rail domain.

In addition to the three system critical situations found earlier, this study also suggests that *Switching*[2] should be included as a system critical driving situation. Switching appears to be a driving situation which is unlike the aforementioned three driving situations. The ethnographic observations that were made in this study were observations of trains that had a route, and which were not classified as a switch. Hence, the basis for classifying switching as a separate system critical situation has been based on interviews, secondary documents, and various incident and accident reports which refer to switching. Interviews with safety officials and train drivers

---

2 In this article switching refers to movements of rolling material with the intention of moving material within a station or a sidetrack area.

Table 3.1 Observable and Non-observable Actions in Train Driving System Critical Situations (adapted from Jansson et al., 2006, p. 44).

Actions	Leaving a station	Out on the route	Approaching a station
Observable Actions	Controlling the platform and the doors to be shut, using the mirror	Weighing speed against comfort depending on whether late or on time	Watching for signals expected to show that the switches are clear
	Supervising, detecting and controlling signals and signs	Supervising, detecting and controlling signals and signs	Supervising, detecting and controlling signals and signs
	Listening to and watching for messages from the ATC system	Listening to and watching for messages from the ATC system	Listening to and watching for messages from the ATC system
	Watching for people and unexpected objects	Watching for people and unexpected objects	Watching for people and unexpected objects
Non-observable actions	Judging time available and preparing for next section	Judging speed ahead in orderto avoid warnings or braking	Calculating braking power and braking distance needed
	Calculating power needed to leave station	Judging time in order to be able to be on time	Preparing to enter the station

support classifying shifting as an independent driving situation which calls for unique use of information and specific actions. Switching is referred to as being a task separate from train driving. However, train drivers are required to master switching, which is referred to as being especially dangerous and demanding (Borgersen, 2001). Switching will be discussed in more detail later in this paper.

4.1.1 Leaving a station

It has been found that when leaving the station the train driver is mainly focused on leaving the station as quickly and safely as possible. When drivers are preparing for leaving, and when they start to leave the station, there is a high level of focus and attention. It is important that the conductor has given the 'safe to leave' signal and that the doors are secured. Drivers make themselves aware of how much time is needed for passengers to board safely, and attempt to leave the station as quickly as possible. Keeping to the timetable is important for the overall flow of the system as well as for the passengers, and drivers attempt to calculate how much power is needed to get the train to accelerate quickly and smoothly. On local routes, the time out on the route where there are opportunities for optimising speed are quite limited. Hence leaving and arriving at stations on time is of the essence and drivers are aware of this when considering the balance between safety and efficiency. Drivers are especially aware of passengers that arrive late, as they pose an extra danger when they approach the train. In addition, late passengers may delay the train's departure.

4.1.2 Out on the route

While on the route, drivers are in a more relaxed cognitive state. They are concerned with keeping to the timetable and calculating the speed needed to reach the next station on time. At the same time, they are aware of the speed limits and wish to keep to these in order to avoid automatic braking by the Automatic Train Control (ATC) system.[3] Drivers experience driving out on the route often as 'driving on green'. In other words, it is expected that they will most likely be able to drive without interruption to the next station. They are, however, aware that they may occasionally encounter signals that require them to either reduce speed or stop. Drivers take the specifications of the specific train and its load into account to calculate the distance it would take to stop with the specific train set when approaching signals. Also, brake tests are often carried out once while out on a route. Usually, this is performed at an early on the route.

4.1.3 Approaching a station

When nearing a station, drivers shift to a more focused, attentive state. They look out of the cab window for signals informing them whether or not it is safe to continue into the station area. Drivers prepare themselves for a potential stop. They calculate speed in relation the braking capacity of the train set and calculate what the effects the weather conditions, local track conditions, and distance to the signals may have on the ability to stop. At certain local stations, there are many lines and many signals in the line of view. Drivers are concerned with locating and reading the signals that apply for their specific line, and ignoring irrelevant signals and signs. In addition, drivers are concerned with approaching the station and making a smooth stop and making themselves aware of the conditions of the platform and what is happening on the platform. Also, drivers estimate the best stop position along the platform in relationship to where passengers are standing/usually stand.

4.1.4 Switching

Switching is the task that perhaps may be considered the most complex and demanding task for drivers. Switching has traditionally been associated with danger and calls for careful attention (Borgersen, 2001). In Norway, switching may be defined as 'movements of rolling material with the intention of moving material

---

3   Automatic Train Control (ATC) is 'the system for automatically controlling train movements and directing train operations' (Railway Technical Web Pages, 2007). The Norwegian infrastructure has some track sections with complete ATC, some track sections with partial ATC, and some sections without ATC coverage.

within a station or a sidetrack area' which occurs without specific orders (Borgersen, 2001). Switching is therefore not classified in the same way as a train. In contrast, a train is defined as 'rolling material which is driven out on the line' and which always is carried out with a specific order. A train may be driven with or without a route. Often, the task of switching has been handled separately from train driving due to the technical differences related to their operation. In other words, 'switching' is not train driving, but is a locomotive driver's task. Train drivers use locomotives to move rolling material, and from a user perspective this can be considered 'driving'. Perhaps 'Driver Operator Tasks' would be a more suitable term. The term 'Driver Operating Tasks' would thus include the tasks drivers are required to operate in the rail domain; train-driving and switching would be covered under the term Driver Operating Tasks.

When drivers are 'a switch' there are many conditions to be considered. The task is often dependant upon the specific context. While switching, it is especially important to be aware of the state of the switching signals, or more specifically, the dwarf signals. These are the signals that apply specifically to a switch in the Norwegian railway system. They show whether switching is permitted, denied, or permitted with caution. Drivers often use handbrakes in this task and the airbrakes are disabled. The demands and actions of the driver are often dependant upon the kind of switching task to be performed, and the context with which it occurs. In general, drivers are in a high state of attention and awareness. They need to be especially aware of how braking affects the material – something which requires special calculations with regard to the specific material that is to be moved. If drivers are to hook up to new material they need to be aware of the speed in relation to safe and smooth contact with the new material and be aware of persons in the immediate vicinity.

In some instances, a train which is on route may act as a switch. This means that in certain situations a train may have to comply with the dwarf signals which normally apply to a switch. Hence, although technically defined as a train, in reality the rolling material acts as a switch. For example, if a dwarf signal belonging to the driver's track shows 'cautious switching allowed,' in practice it means there is an upcoming dwarf signal that shows 'switching not allowed.' If a train meets a 'switching not allowed signal' it will be required to stop in most cases (even though the signal is a *switching* signal). In effect, the signal shows the train driver that entry to the station is temporarily delayed. Drivers must be aware of the meaning of the different dwarf signals theoretically and at the same time evaluate which driver the 'cautious shifting allowed' signal actually refers to in practice.

4.1.5 Preliminary summary

Preliminary analysis has indicated four system critical situations. Switching is suggested to be unique driving situation which is quite complex and demanding. In future studies, a more detailed task analysis of switching as a work task for drivers (including ethnographic observations) should be considered. It is suggested that special attention should be paid to how dwarf signals function in theory (for a shift) and how they function in practice (relevance for a switch *and* trains en route).

Local train driving has been found to be characterised by more or less systematic variations in concentration and attention intensity. Highly demanding situations (approaching a station, leaving a station and switching) are alleviated by less demanding periods (out on route). Local train routes, especially inner city routes, have relatively short routes with many stops. The cognitive demand thus changes quite often, and quite dramatically, throughout the route. In the future, the effects of high-demand and low-demand situations in local, inner-city, rural, and inter-city routes should be looked at. It appears that the train driving task differs somewhat, if not fundamentally, with various kinds of routes, due to the difference in time intervals between high- and low-demand situations. For example, inter-city routes have a much longer *out on route* period. This could, perhaps, have important implications for the Driver Operating Task in itself and also the support system required for each type of task.

Jansson, et al. (2006, p. 45) conclude that 'drivers' work can be divided into three rather different time intervals: a long-range interval with an interaction between the train and a rather distant environment; a short-term interval, with an interaction between the train cab and the visible surroundings; and finally, an immediate sense interval with an interaction mainly in terms of braking and feed-back from the stopping train'. Our findings are in accordance with those of Jansson et al. However, it is important to note that in actual driving situations, the interaction described within the different time intervals appears to occur more or less simultaneously. On the one hand, drivers are aware of the immediate sense interval (they 'feel' the train set's reaction to braking or slowing down), while at the same time considering the visible surroundings (looking for moose trackside), and attempting to plan for the future (thinking about where to expect upcoming signals). So, while it may be helpful to consider the different time intervals as separate in some situations, it is important to keep in mind that in actual driving situations the driver is simultaneously interacting with the system at other time intervals.

4.2 Safety Evaluations

The following section attempts to evaluate human factors issues relevant to the train driving task with the aim of identifying some areas which may be improved in the future. In modern human factors research it has been considered important to monitor and evaluate safety issues continuously. It is worth mentioning that modern rules and regulations call for continuous auditing and evaluation of system safety control. Such regulations came into effect on 1 January 2007 (FOR 2006-12-06 nr 1356). In addition, a new law for work environments was enacted during the course of this project (1 January 2006) which has ramifications for diverse railway safety procedures (LOV 2005-06-17).

The second part of this paper is organised according to order of importance, with continual discussion on each point. The human factors related aspects presented and discussed in the following are: Infrastructure; Incident and Accident Reporting and Follow-up; In-Cab Man-Machine-Interaction (MMI); Situational Awareness; and Training and Experience.

4.2.1 Infrastructure

The Norwegian environment offers a number of challenges for the rail system. This is especially relevant when the immediate track environment is considered. In Norway there are dense forests, high and steep mountains, fjords, a large number of lakes, and other waterlogged areas. In addition, weather conditions such as heavy snow, strong winds and heavy rainfall create challenges for the infrastructure and the train driving task. The Norwegian environmental and weather-related conditions are demanding for those who are responsible for designing the rail infrastructure, those who keep the infrastructure maintained, and those who operate it. Many routes have overhanging foliage over curved tracks, creating a 'tunnel-like' driving experience. Track crossings occur in a variety of settings and are handled in a number of different ways. All of these factors contribute to making the driver's ability to plan ahead difficult and highly dependant on experience.

It is well-known in the research literature that aspects related to infrastructure have important implications for the train driving task. Safe train driving presupposes a safe, well-functioning, and supportive infrastructure. The infrastructure is a major contributor to the behaviour shaping constraints related to train driving. If the conditions of the track makes it almost impossible to brake, for example due to ice on the track, the driver will have to take this into consideration when assessing how to approach his or her task. To be able to take track conditions into consideration, a driver must be able to have access to information regarding the tracks condition in any given location.

4.2.1.1 Driver-Infrastructure Interaction

A driver can only stop at required points along the track if the signal system is functioning, relays information in a way in which drivers can understand, and relays important information timely. This has been made clear in recent accidents where trains have collided with rock slides and snow avalanches (Accident Investigation Board Norway), and also in SPAD incidents (Norwegian State Railways). If drivers are unable to detect slides which have occurred on the track ahead quickly (perhaps simultaneously as they occur), accidents will be unavoidable.

There are several aspects of the Norwegian rail infrastructure which are of concern from a human factors point of view. The current investigations have revealed great variation in the execution of train driving procedures – something that is quite often the result of local infrastructure conditions. The more context dependant the situation, the more dependant safety is upon individual actors and single actions. A situation which is highly context dependant will result in a greater degree of performance variation, which often is difficult to predict. In highly context-dependent situations, agents will have a rather low degree of control (Hollnagel, 1998). Their actions will be more dependent upon 'here and now' decisions and the total system safety will be more reliant upon individual's ability to make appropriate decisions and/or take appropriate actions. If one considers the time intervals drivers interact with the system mentioned earlier (immediate, short-term, and long-range), the ability to work long-range will be considerably more difficult when the system is highly context dependant.

4.2.1.2 Placement and condition of signals and signs

Signals and signs are often placed in areas which are either hard to see (for example, beyond a bend) or in positions which are exceptions to the rule. Regulations call for placement of signals as a rule to be in areas to the right of, or above, the relevant line (FOR 2001-4-12 nr. 1336). There are instructions referring to how the placement of signals should be handled in special cases. However, there are a vast number of cases where signals are placed in positions which are exceptions to the rule.

In addition, the conditions of signals and signs are often such that they are difficult or impossible to read. Some signals may be dirty or snow covered, while others may be dysfunctional or out of order.

Some of the infrastructural conditions which are less than optimal remain in this state for some time, while others are daily (if not hourly) exceptions. Variation in the long-term placement and condition of signals creates a situation which demands a high degree of local knowledge for users (drivers). In addition, there are frequently daily announcements and notifications regarding the placement and condition of the infrastructure on any given route, which drivers must be familiar with before starting his/her shift. Due to variations in the infrastructure (such as dysfunctional signals, work on platform, and the state of ATC) it is quite often necessary for drivers to make minor or major exceptions to the rules when carrying out procedures. Performance is thus highly dependant upon the ability of individual drivers to integrate new information into their actions when driving.

Our investigations suggest that there may be problems tied to the frequency and numbers of announcements and notifications drivers are presented with, and also to the time available to become acquainted with and integrate the information given. To our knowledge, there have been no studies focussing on this problem in Norway. It is therefore uncertain as to how much time is actually necessary to be able to read, become acquainted with, and to integrate new information presented in announcements and notifications. Most likely, the time needed varies greatly according to the number and type of changes that are being announced, and according to the language used in the documents. In addition, the amount of time needed to integrate new information is likely to vary from individual to individual.

4.2.1.3 Light signal system

Several aspects of the light signal system were of concern from a user point of view. Light signals were found to have different purposes and relevance according to the specific location and context. More specifically, the Norwegian 'dwarf' signals were found to be especially context dependant.

In addition, it is not unproblematic from a user point of view that signs and signals are physically placed at different trackside locations. Although the placement of signs and signals follow a general pattern determined by regulations, it appears that there are a considerable number of exceptions to the rule. Most signals are placed on the right side of the track, but sometimes signals are found on the left side and also above the track. It may be that drivers' reaction times may vary according to whether signals are shown at different trackside locations.

General concerns with being able to see relevant signs and signals were voiced in interviews. 'Dwarf' signals were often voiced as being difficult to read. Drivers claimed to have experienced signal failure or malfunction as a large source of stress. They also have expressed that at times they find inconsistencies and incoherence in the infrastructure to be stressful. Changes and deviations from the general rule in the infrastructure put a greater demand on the operator (the train driver) and are experienced by drivers as stressful.

4.2.1.4 Preliminary summary of infrastructure analysis

Infrastructure shapes the working constraints of train drivers. The more complex and unpredictable the infrastructure, the less control the user (driver) will have (Hollnagel, 1998). Complexity and unpredictability contribute to creating a less stable control situation. Actors' (drivers') ability to plan ahead and act rationally are reduced, and are more dependent upon the immediate conditions and situations experienced, something which is regarded to as context dependency.

To conclude, there were several aspects regarding the Norwegian rail infrastructure which are concerning, and which represent weaknesses in the rail system as a whole. In particular, a high degree of context dependency was found. While context independent signing is independent of the operator, context dependant signing is operator dependant (Flach, 1995). Context dependant signs and signals increase the amount cognitive work needed to complete the task safely and smoothly, demands route experience from drivers, and restricts optimal situational awareness (SA). Context-independent signs, signals, and other aspects of the infrastructure should be emphasised and developed in the future.

4.2.2 Incident and Accident Reporting and Follow-up

Often, investigations of accidents or potentially dangerous incidents are used to uncover weaknesses in complex socio-technical systems, and are required by law. Studying incidents of a potentially dangerous character is an economical methodology that can enlighten researchers and practitioners as to the vulnerable parts of the train system (Kirwan & Ainsworth, 1992). The nuclear and aero industries have long experience with reporting and analysis of incidents and accidents, with positive results. Fruitful analysis is dependant upon a high quality reporting system. Reports of near-misses and incidents of lesser degrees of severity can often be used to guide safety prevention work. Such cases are usually closer to normal work-task activities and in this respect should they should be paid attention. Incidents and accident types (rather than specific cases or episodes) indicate weaknesses in the socio-technical system and can thus pinpoint which areas need to be focused upon (Rasmussen et al., 1990). However, to be able to learn from reporting, it is necessary that investigation has a stop rule which goes beyond the 'human error' and which aims to find the latent causes leading up to the unwanted event (Rasmussen, 1990). To maintain or improve system safety, the aim of reporting and report analysis should be to identify and understand missing and/or insufficient barriers rather than finding human error (Hollnagel 2004; Rasmussen, 1990).

4.2.2.1 Reporting

Procedures related to the reporting of SPAD incidents in Norwegian State Railways were examined and found to vary. The reporting quality and practice was seen to be somewhat dependant upon the person carrying out the investigation, with no established practice of anonymous reporting internally within the company. Drivers are required, however, to report sub-optimal conditions and unwanted events. It is the drivers' closest leader who has the main responsibility to evaluate an incident and decide how it should be reported. A guide (checklist) for report writing has been developed in the past, but in practice this checklist is in used very seldom (if at all). The enterprise logs reported incidents and accidents using SYNERGI-brand risk-management software, but lacked formal guidelines for report writing and useful tools with which to analyse this data from a systems perspective.

Recently, the Norwegian State Railways has established that a SPAD event is one of the most potentially hazardous situations in the Norwegian railway system. This type of event has been evaluated as occurring at an uncomfortably high rate, and is regarded as an unwanted incident. The reporting of serious incidents and accidents is controlled by governmental laws and regulations, whilst the reporting practices and investigation of less serious incidents or potential incidents/near-misses are left to the enterprises themselves.

To date, there has been little international research using the systemic approach, and even less national research concerning the reporting practices and analysis methods used by safety managers to understand SPAD events. There have, however, been several international publications regarding human factors and reporting systems in general, as well as publications on understanding violations and safety cultures in the rail domain (see Kecklund, 2001; Wilson & Norris, 2005; 2006; Wilson, Norris, Clark & Mills, 2005)

CREAM is a standardised tool which can be used to guide report writing and analysis. It may be applied both retrospectively, using existing incident/accident reports, and prospectively as a form for second generation human reliability assessment (HRA) (Hollnagel, 1998). It is designed to be used in analysis of complex social-technical systems, making it a potential methodology to be used in the rail domain (Hollnagel, 1983; Hollnagel, 2005a). CREAM is perhaps the latest complete analysis methodology which acknowledges the important role that context plays in human performance variability. In CREAM, contextual information is important when considering the specific actions of individual operators in specific incidents.

Table 3.2 Human Factors information left blank in Norwegian State Railways Synergi Reports.

Who	External client involved Comments, suggested cause, suggested Initiatives Shift Weekday Experience in position (months)
Classification	Guiding documentation – rules Involved Equipment/system Equipment/System description Involved Infrastructure Internal Time
Initiative	Another initiative responsible person E-mail – another initiative responsible person

Table 3.3 Human Factors Issues not addressed in Norwegian State Railways Synergi reports.

Organisational Factors	The quality of the roles and responsibilities of team members Additional Support Communication Systems Safety Management System Instructions and guidelines for externally oriented activities Role of External agencies, etc. Quality of driver–train traffic manager communications Crew collaboration
Working Conditions	Ambient lighting Glare on screens Noise from alarms Interruptions from the task Quality of interaction with in-cab interface Weather conditions as experienced by driver
Training and Experience	Familiarity with particular train set Experience with task Experience with route (months) General experience as driver (months) Experience with new technology Age of driver Driver's involvement with previous incidents

Since it is important to consider reporting together with report analysis methodology to ensure compatibility and usability, Norwegian State Railways' SPAD reports were seen in the light of CREAM, and the domain-specific DREAM (Driver Reliability and Error analysis) method (Hollnagel, 1998; Ljung, Furberg & Hollnagel, 2004)

SPAD reports have been found to lack important information needed for modern human factors system analyses. Table 3.2 shows the human factors related issues that were often unaccounted for in Synergi reports, while Table 3.3 lists human factors issues that are important to consider in modern accident investigation methodologies but that were not addressed in Synergi reports.

SPAD reports were often found to refer to specific problems with the infrastructure as causes. Reports which find infrastructural problems are often constructively sent to those persons who are responsible for infrastructure. Often, problems were related to signal malfunction and tied to certain 'problem locations'.

Person-related causes (lack of attention, lack of following procedure) were also found to be quite frequent, but such causes are more difficult to work with constructively. As a rule, it is more constructive to dig deeper to understand the latent causes: What happened before the driver 'missed' a stop signal? Was the individual exhausted due to lack of sleep? Did the driver lose out on sleep because of the psychosocial environment at work? Were there a great number of signal malfunctions or was trackside work going on? Was the air conditioning making so much noise that it led to fatigue? Also, is it the case that 'failure to follow procedures' is used as a 'cause' in situations where normal work adaptation fails (thus being unacceptable), while in other situations adjustments to procedures is regarded as regular practice (and accepted or even encouraged)?

The behaviour of users in unfamiliar or unexpected events is often conditioned in everyday, normal work (Rasmussen et al., 1990). As mentioned earlier, the normal train driving task in Norway is characterised by a high degree of context dependency and often reliant upon drivers' ability to cope with variation. Often, exceptions to the rules are necessary to keep the system operating. Future research may indicate whether it is the referrals to 'lack of attention' or 'failure to follow procedures' in incident reports that indicate an area of problematic interaction between driver and infrastructure.

4.2.2.2 Follow-up

The following up of incidents, such as reporting practice and quality, seems to vary. Some incidents are followed up by a talk or interview with the involved driver, some are followed up with in-cab coaching or evaluation, while others are followed up by time spent in a train simulator.

While many of the decisions related to follow-up of drivers are certainly sufficient and of a good quality, this investigation has uncovered incidents where the treatment of drivers is quite alarming. This can be shown by the following example.

One driver, who had been involved in several incidents involving person-related factors, had received orders to undertake a special training course and simulator training. There were a number of aspects related to how this individual was treated that are of concern. The situation involving simulator training was perhaps the most alarming. The individual described the situation as being a 'test' where he was evaluated by several of his superiors simultaneously. He described the experience as 'traumatising' and said that he was 'scared and nervous.' He reported that the simulation he was made to drive included an abnormal number of critical situations which required special attention. The driving situation which was simulated would most likely never occur in actual train driving, and the driver reported entering into a kind of 'stand-still' or 'mind blackout' due to the stress and abnormality of the whole situation. Physical symptoms of stress related to the autonomic nervous system, such as sweating and heart racing, were reported. After the simulated driving was finished, the individual was called into an interview, where the results of his simulated driving were the main topic of discussion. There, all of his superiors had the opportunity to ask questions and comment on his simulated driving task. The individual felt the interview was like an 'interrogation', and that it was unfair. He reported feeling this due to the abnormality of the scene he was asked to simulate driving, and the stressed state of mind he was physically and mentally in at the time of the simulated driving due to the nature of the situation. This particular individual appeared to be aware of his general special need to have 'actual driving experience' in order to understand how to drive appropriately in certain situations – something which his driving instructor had mentioned during his initial training as a train driver. The training coach (the driver who had responsibility to accompany the individual while still under training) stopped being a coach. This was reported as being due to the fact that the issues the training coach pointed out to safety officials (about the individual under examination and other individuals) repeatedly were either ignored or not taken seriously.

Although this is a description of one individual, it indicates the need for a deeper understanding or investigation of the techniques used to follow up drivers involved in unwanted incidents. It is undoubtedly the enterprises' responsibility to identify those persons which are incapable of becoming train drivers. Regardless, the process of evaluation and follow-up should not be carried out in a manner which is detrimental to individuals' physical or mental health and well-being – which appears to have been the case in the example described.

4.2.2.3 Preliminary summary of incident and accident report analysis

The analysis of SPAD incident reports and systems analysis revealed several weaknesses involved with the reporting practice, report quality, and follow-up routines.

There should be clear instructions as to when and how reporting should be done, and the stop-rule should reflect the wish to identify latent causes in addition to direct causes. Inattention and/or lack of concentration and failure to follow procedures' should not be regarded as sufficient latent causes. We suggest a stop-rule which attempts to find factors contributing to inattention and lack of following procedures. In addition, a system which allows for and encourages anonymous reporting of near-accidents and problematic system-interaction concerns should be developed and implemented. This system should be user-friendly to encourage its use.

It is recommended that a guide or checklist is used actively when incident investigations are initiated, and that a high-quality semi-structured interview is performed with the train driver to ensure that the details needed to perform modern human factors incident/accident analysis are obtained shortly after the incident occurs.

The Cognitive Interview (CI) is a specific interview technique which may be used in incident/accident investigation (Memon, 1999). The CI has been used previously in forensic settings to increase the effectiveness of communication and to improve witness performance by drawing upon cognitive and social psychology to improve interviewer–interviewee relations and memory retrieval. A person conducting a CI will attempt to establish a trusting relationship with the interviewee and assist the interviewee on a mental journey back to the physical scene and their mental state at the time of the incident. The interviewer will encourage the interviewee to report everything that comes to mind, regardless of how important the interviewee regards the information. Allowing the interviewee time to respond and think is important. Questions should be limited and asked timely. Also, the interviewee should be asked to describe the situation or event in reverse order and/or from another perspective.

To our knowledge, the CI has not been used in human factors or rail investigation previously, though it is a very relevant method to use. The CI technique has been used in the course of this project, and preliminary results suggest that the CI may prove to be a very valuable technique. It is designed to increase contextual and experiential information in witness reports, and is a methodology that can be used by practitioners or non-psychologists/cognitive agronomists. It is likely that the managers closest to the drivers would be able to conduct such interviews without any special training in cognitive ergonomics or psychology-based interview techniques. A specific interview technique may prove to increase the quality and amount of information currently collected from incident/accident interviews. The CI has been shown to exhibit more correct information than the standardised police interviews without increasing incorrect or confabulated information (Geiselman, Fisher, MacKinnon & Holland, 1985; Fisher, Geiselman & Amador, 1989, cited in Memon, 1999). Also, the CI has been seen as superior to other recognised interview techniques, such as the structured interview, when it comes to obtaining more information without increasing the amount of incorrect/confabulated information – especially when it is possible to corroborate with other information sources (Memon, 1999). Further research on the use of CI in system safety work may provide practitioners with information about the use of the CI technique in the rail industry. The goal should be, at any rate, for practitioners to find a useable, high quality, psychologically based interview technique to be used in accident investigation.

Follow-up of initial training and of individuals who are involved in incidents and accidents should be handled supportively and with the aim of reducing the risk of repetition. Investigations may reveal system interaction problems and/or special needs of individual drivers. It is important to keep in mind that most incidents are the products of problems with system interaction, and should be looked upon as opportunities with which to find insufficient or missing barriers in the system.

Table 3.4 Common Performance Conditions in the Railway: CPC explanations.

CPC EXPLANATIONS		
Common Performance Conditions – The Scene		
CPC	**PARAMETER**	**EXPLANATION/DESCRIPTION**
Traffic Environment's shaping factors/ frame	Type of Traffic environment	Describe the traffic environment Urban or rural routes? What kind of foliage?
	Complexity	How complex was the traffic environment? How many tracks? Single track or double? Main track or side track? Single track? Approaching stopping Station? Meeting a train? Road crossings? Were there any tunnels or bends in the track in the area? Were there many signals/signs?
	Information	Were the signals clear and easy to see and be understood? Were the track signs clear? Is information given with sufficient time so that the driver can prevent an accident? Is any important information missing?
Driving Conditions	Traffic Density	How many other trains were in the area? Low volume or rush traffic? Are there any statistics for when there is rush hour traffic in this specific area/route?
	Track conditions/friction	Where there any factors, such as weather-related factors Dry and above freezing = optimal Dry and below freezing = good Wet track and above freezing = good Wet track and below freezing = tolerable Snow on track = Tolerable/Bad Ice on Track = Bad
	Visibility – Weather and Lighting	How was the weather influencing the visibility? Was the visual field clear? Was it influenced by the dark, snow or rain, reflecting sun, etc.? If applicable, was/were artificial lighting/lamps working and sufficient?
	Visibility – Obstructing Objects	Was the visibility clear, or obstructed by poles, bushes, or other physical objects? Were there any blind spots?
	Infrastructure	Railway signalling installation: track conditions, lighting, signals, signs, platform. Section with remote control, with or without line block? Main track, side track, single or double track section? Station with porter, remote-controlled station, or boarder station Section with Full Automatic Train Control (FATC) or Part Automatic Train Control (PATC) Was the track in optimal conditions? Was there any track-side work being performed in the area of the incident/accident? When applicable, was the lighting as it should be? Was it optimal for the weather conditions? Were signals in working condition and easy to be seen/read? Were they placed beyond a bend, for example? Were signs placed where they should be and easy to be seen/read? If a signal was lit, did the driver have sufficient time to react? Any sudden changes in the signaling? Were there any items on the platform that could be distracting (items that normally aren't there)? Was there anything out of the ordinary happening on the platform?

Common Performance Conditions – The Driver		
CPC	PARAMETER	EXPLANATION/DESCRIPTION
Driver's working environment/ conditions for work	MMI – Interface Only	Are the instruments and in optimal working order? MMI- How are the diverse interfaces designed? Are they user-friendly? Is the interaction designed with safety in mind?
	MMI – Combination of interface	MMI – Is the combination of interfaces beneficial for the driver, and could it cause a problem? Is the interaction designed with safety in mind?
	In-cab working environment	Was there anything in the cab environment that could be experienced as uncomfortable or stressful? For example, was the seat in working order and easy to adjust? Were there any disturbing or distracting noises or sounds such as a loud air conditioning or window washing fluid?
Driver's presuppositions	Time of Day/Day of Week	A well-known psychological phenomenon is that one acts outside of the normal daytime rhythm, in other words outside of the time when one normally sleeps, one has a reduced performance stability. This is known as effects of circadian rhythm. Did the incident/accident take place within the driver's normal sleep-wake rhythm or not? State preferred normal sleep-wake rhythm, if the person has worked at night, been up late, etc. How did the driver, or others involved, sleep during the night(s) before the incident/accident? Regarding shift work, how has the individual worked recently?
	Number of Simultaneous Activities/Goals	How many activities did the driver have at the same time? For example, checking watch to regulate progress against timetable, braking, and checking platform. Was he/she talking on a mobile phone, looking/listening for some kind of information (give examples), adjusting seat or heating/air conditioning, hearing a distracting noise from the cabin area? Here, time, or lack of it, is important. Did the driver have enough time to see the signals and react? For example, was he or she allowed time sufficient to both see the stop signal and also brake in time? Or did the signal 'sneak up' on the driver?
	Driver's physical and psychological health	How was the driver feeling? Feeling well, or in reduced health in any way? Was the driver uncomfortably warm/cold? Flu or cold symptoms? Were there any social factors that could have influenced the individual? For example, works social environment, conflict with co-workers, or family situation? Which aspects to be reported may vary from case to case? For example, low blood sugar levels due to a long period since the last meal.
	Speed in relation to speed limits	Did the driver keep to the speed limits or did he/she drive too fast/slow? If so, how much?

Common Performance Conditions – The Driver		
CPC	PARAMETER	EXPLANATION/DESCRIPTION
Driver's Experience and Training/Education	Driving Habits	Where was the driver used to driving? Number of years as train driver? Total km per year? Does the driver drive in all kinds of environments, or is his/her driving restricted or reduced to specific areas? Any specific reasoning for this?
	Driver's acquaintance with the traffic environment	Was this the first time the driver was at this location? If not, how often does the driver usually drive the area of the incident/accident?
	Driver's acquaintance with the train set	Is this a new train set? If not, how often does the driver drive this specific train set? Have there been any recent changes in the cabin or driving area which the driver is unacquainted with? How does the driver subjectively feel that this particular train set is to drive? Is it comfortable and easy to drive?
	Driver's training/education	There may be special regulations or procedures for a variety of situations (ex..) If so, is the driver educated in these, and are they applicable in the traffic situation in which the driver found him-/herself? Are the plans, regulations and/or procedures easy to access? If a special train set is involved, is the driver educated in the use of such, and how extensive is his/her level of experience?

Common Performance Conditions – The Organisation		
CPC	PARAMETER	EXPLANATION/DESCRIPTION
Adequacy of Organisation	Roles and Responsibilities of team members	Are the roles and responsibilities of team members clearly defined and non-conflicting? Are there any time/efficiency/safety conflicts? If so, elaborate. For example, have drivers had sufficient time to read through the daily notices about route modifications, changes; notices regarding special circumstances for the route?
	Communication	Were messages/orders given/received when, how and where they should be? Were there any misunderstandings?
	Role of external agencies	Were there any external agencies involved? For example. outsourcing of snow clearing, etc. If so, were their roles and responsibilities clearly defined? Were they educated in the rules/regulations that apply? What kind of experience does the agency have?

Table 3.5 Common Performance Conditions in the Railway: CPC form.

Common Performance Conditions Form					
Common Performance Conditions – The Scene					
CPC	PARAMETER	EVALUATION	COMMENTS	INFO-SOURSE	INFO-QUALITY
Traffic Environment's shaping factors/ frame	Type of Traffic environment	Rural Urban			
	Complexity	Little complexity Moderately complex			
	Information	Supporting Approved Tolerable Inadequate			
Driving Conditions	Traffic Density	Little traffic volume Steady flow Rush traffic			
	Track conditions/friction	Optimal Good Tolerable Bad			
	Visibility – Weather and Lighting	Optimal Good Tolerable Bad			
	Visibility – Obstructing Objects	Optimal Good Tolerable Bad			
	Infrastructure	Supporting Approved Tolerable Inadequate			

Common Performance Conditions – The Driver					
CPC	PARAMETER	EVALUATION	COMMENTS	INFO-SOURSE	INFO-QUALITY
Driver's working environment/ conditions for work	MMI – Interface Only	Supporting Approved Tolerable Inadequate			
	MMI – Combination of interface	Supporting Approved Tolerable Inadequate			
	In-cab working environment	Supporting Approved Tolerable Inadequate			
	Time of Day/Day of Week	In daytime rhythm? In between/both Outside of daytime rhythm			

Common Performance Conditions – The Driver					
CPC	PARAMETER	EVALUATION	COMMENTS	INFO-SOURSE	INFO-QUALITY
Driver's presuppositions	Number of Simultaneous Activities/Goals	Less than capacity Matching capacity More than capacity			
	Driver's physical and psychological health	Good Reduced			
	Speed in relation to speed limits	Over Same Under Much Under			
	Driving Habits	Sufficient, comprehensive/ extensive Sufficient, but limited Insufficient			
	Driver's acquaintance with the traffic environment	Passes daily Driver there many times before Driven there sometimes before Never driven there before			
	Driver's acquaintance with the train set	Sufficient, comprehensive/ extensive Sufficient, but limited Insufficient			
	Driver's training/education	Supporting Approved Tolerable Inadequate			

Common Performance Conditions – The Organisation					
CPC	PARAMETER	EVALUATION	COMMENTS	INFO-SOURCE	INFO-QUALITY
Adequacy of Organisation	Roles and responsibilities of team members	Clear and precise Some discrepancies Many discrepancies			
	Communication	Supporting Approved Tolerable Inadequate			
	Role of external agencies	Supporting Approved Tolerable Inadequate			

During the course of this project, a guide has been developed for train incident and accident reporting, based upon modern cognitive ergonomics human risk analysis methods. The guide has primarily been based upon CREAM and DREAM (Driver Reliability Error Analysis Methodology (Hollnagel, 1998; Ljung et al., 2004). However, the guide has also been based upon other relevant modern rail (and other) human factors research, in addition to the knowledge gained throughout the course of this project (see for example, Lawton & Ward, 2005; Shepherd & Marshall, 2005; Wilson & Norris, 2006; 2005; Wilson et al., 2005). The guide includes parameters and descriptions of the common performance conditions related to the scene, the driver, and the organisation. The guide is designed to be used by persons with or without a background in human factors. Persons may use the guide when interviewing persons involved with an incident/accident, in addition to report writing. Table 3.4 shows an explanation of the *Common Performance Conditions* (Hollnagel, 1998), and their parameters. Table 3.5 shows a form which can be used to assist the data collection process and subsequent analysis.

The guide presented in Table 3.4 and Table 3.5 is designed for use in all incident and accident investigations. By investigating incidents of a lesser degree more thoroughly, it may be possible to uncover weaknesses in the system that most likely could contribute to serious incidents or accidents in the future. It is likely that there will be a much higher number of less serious incidents to investigate, making it easier to detect problematic system interaction and incident 'types.' However, to identify the latent causes of minor or 'non-serious' incidents, it is necessary to have the same high quality information as is required for serious incidents and accidents.

4.2.3 In-Cab Man-Machine-Interaction (MMI)

Distractions and task interruptions, together with, for example, technical failure and unexpected events may be a fatal combination (Reason, 1990). This is referred to as an *interaction effect*. It is generally important to minimise aspects which may contribute to such an effect. When the task performed in normal conditions is moderately to highly demanding or difficult, then the likelihood that an interaction effect will produce an unwanted outcome rises. Train driving situations, such as approaching a station, leaving a station, and shifting are examples of normal activity which are cognitively demanding. Technical failures or local abnormalities or deviations from the norm also contribute to enhancing the cognitive demand of the train driving situation. Since the train driving task in local routes is often cognitively demanding, it is important that the in-cab environment is as supportive as possible.

Several observations in the in-cab area revealed objects or incidences which represented safety hazards, such as very noisy window wipers, knobs on the interface that fell off during use, noisy ventilation systems, dysfunctional seats, strong smells in the cab, annoyingly bright screens on the interface, and dirty rear view mirrors. These could be hazardous by themselves and could potentially contribute to an interaction effect, in which case they should be eliminated or minimised. Of the aforementioned observed matters which represented the most danger, but were the easiest and cheapest to fix, were dirty mirrors and loose knobs on the interface. Mirrors and knobs are very important artefacts in the in-cab environment. Dirty mirrors and/or windows obstruct drivers' ability to see clearly. One of the most obvious aspects of the train driving task is the importance of visual cues. Drivers use visual cues continuously in their work to drive safely, and it is absolutely vital that they can see clearly out of all windows and use all mirrors.

Regarding loose knobs, for the most part the danger is tied to the function the knob represents. According to Petersen (2004, p. 266, original italics), 'it is important to consider how changes in the controlled system and its environment influence the *control situation*: the possibility of bringing about system state changes by performing control actions on the controlled system (*the control possibilities*), and the requirements for bringing about appropriate state changes in the controlled system (*the control requirements*)'. Each knob represents potential change that may be brought about in the controlled system (rail system). Lack of knob (the input device) hinders the actuation of the effector mechanisms. If the knob falls off, the ability to bring about the specific change represented by that knob will be hindered or eliminated. The control situation is instantly changed, by limiting the changes in the system that can be brought about by turning the knob (control possibilities). The direct danger depends, of course, upon the function represented by the knob and the vitality of that function for safe travel. The indirect danger is tied to the reaction of the driver. While one driver may lean down to search for the knob while out driving on a route, another may decide to wait until the train has stopped at the next station, while yet another may decide to make an exceptional stop to find and fix the knob. What the driver decides to do with the dysfunctional knob (control action) changes the system state further, creating a control situation with other possibilities and restrictions.

4.2.3.1 Preliminary summary of In-cab MMI analysis

Distracting noises, dysfunctional seats and strong smells in the immediate in cab environment are likely to result in a trial and error method to minimise or eliminate any adverse consequences. Well-documented techniques relating to train driving are the tendency to do what is referred to as 'system tailoring' and 'task tailoring'. System tailoring refers to the fact that operators have a tendency to set up a system in their own way, in a way which has not been considered standard use. If, for example, the noise of the ventilation system is experienced as unpleasant and annoying, drivers will try to minimise this by perhaps turning on and off the ventilation system to get blocks of time without noise. This behaviour, in turn, results in temperature swings in the in-cab environment which may lead to fatigue, in turn affecting the driver's ability to concentrate and remain attentive. In addition, the behaviour itself (turning the switch which regulates the 'off' or 'on' function) presents a danger in that it draws the drivers attention and actions away from the main task at hand (train driving) and averts it to the task of maintaining the temperature.

A more complex issue which should be looked at in the future relates to the different interfaces drivers are required to use. Drivers are to be able to switch between several train sets, all of which have their (more or less) unique interface. Our experience is that the interfaces are somewhat similar, but that each train set has its own distinctive characteristics which require a certain degree of orientation and/or experience.

Human factors research suggests that there may be some adverse effects related to switching between different interfaces. It is widely accepted that the technical system has a tendency to structure the task at hand as such, something regarded to in the literature as 'task tailoring'(Cook & Woods, 1996). The implication is that people will have a tendency to relate to the task *as it is represented through the technical interface*. This is also known as the intention-function problem. It is assumed that the more experience someone has with the interface, the less they are vulnerable to the problem of task tailoring. It would therefore follow that train drivers will relate to the train driving task in one way while driving one train set, and in another way when driving another train set.

Future studies should explore the effects of changing between train sets. Newer train sets often have more advanced technology and different ways of relating to the ATC. It may be that the differences in the interface change the work task in ways that create problems for the drivers (for example, making it take longer to find relevant knobs or to brake smoothly). On the other hand, individuals who have work tasks that are both highly demanding and presented through an interface may benefit from having more hands-on, manual experience (Charlton, 1996). At this point, it is difficult to predict the effect of varying train sets on system safety.

4.2.4 Situational Awareness

Situational awareness (SA) is important for all users of a technical system (Endsley & Garland, 2000; Endsley, Bolte & Jones, 2003). [4] SA is knowledge that operators build up over time, and which is used to predict future situations or events. Often that knowledge provides the basis for decision-making processes. The train driving task, especially the local train routes, is similar to the task of manoeuvring an airplane or fast-speed patrol boat in that the task is highly dynamic. In such situations, the drivers' ability to act upon assumptions about the future is necessary to achieve a good result. In scientific terms these are called proactive actions. Proactive actions are considered to be dependant upon a high degree of situational awareness.

---

4 'Situational awareness is the perception of the elements in the environment within a volume of time and space, the comprehension of their meaning, and the projection of their status in the near future' (Endsley, 1988).

4.2.4.1 System coherence and correspondence

To be able to predict future events, it is necessary for the system to have a high degree of coherence and correspondence (Vicente, 1999). Lack of coherence and correspondence in a system can be a source for human performance variance. A number of inconsistencies were observed in the particular system under study.

There appear to be, at times, quite large discrepancies between procedures and practice. Due to the fact that the actual infrastructure varied quite often, carrying out procedures smoothly and correctly was often seen as being linked to the driver's local knowledge and experience (experience will be discussed later in this report). Much of the train driving task was found to be highly context dependant. It is likely that the inconsistencies in the carrying out of procedures vary due to inconsistencies and lack of correspondence in the infrastructure. There appear to be many exceptions to the rule regarding placement and meaning of signals and signs – something which demands a great deal of local knowledge. Understanding how incoherencies and lack of correspondence in infrastructure affect the task procedures should be considered in the future.

At present, the system conditions place a great cognitive demand on drivers. The more context dependant the situation is, the more the safety of the system is dependant upon the operator. It is important for complex technical systems to work to *reduce* the context dependency of the system as to ensure reliable safety barriers (Flach, 1995). The more the system is dependent upon the operators' understanding and interpretation of the immediate environment, a kind of right 'here right now' situation, the more difficult it becomes to think and act proactively. Drivers will experience a lesser degree of control. There is less room (often related to time) for misjudgements, lack of attention, and interruptions. As a consequence, individual differences (performance variation) will have a larger effect on the total systems safety.

4.2.4.2 Preliminary summary of situation awareness analysis

This project reveals that the current rail system is such that drivers have a relatively low degree of situational awareness. There appear to be many conditions and situations which reduce the drivers' ability to plan ahead. There may be several ways to improve situational awareness by making the rail system less context dependent. Most likely, a variety of initiatives at different levels of the M-T-O triad will be the most successful. A more complete systems analysis of the domain could provide a basis for the design of new technology and other initiatives.

Some ways to make the infrastructure less context dependent have already been discussed in this paper. There may also be technical means of representing upcoming information (state of signal) in the cab, making proactive thinking and actions easier and more accessible. Jansson et al. (2006) attempted to improve existing train driver interfaces using modern human factors techniques. The resulting interface presented the most important information on the current status of the train set, while at the same time presented various aspects of the upcoming environment (including special conditions, future speed profile). In addition, the interface included information which would be useful in the case of an unwanted event or accident, and which functions as a barrier to reduce the ramifications of such events should they first occur.

It may also be possible to increase situational awareness through specific training initiatives and through hands-on experience. Training and Experience will be discussed in the next section.

4.2.5 Training and Experience

When it comes to education and training of actors working in domains where safety is an issue, it is important to have continual evaluations of both education and the types of experience that are necessary. It has not been the aim of this project to make any concrete evaluations of the education train drivers receive. Rather, we have chosen here to focus on drivers who have gone through the required training process, and discuss training and experience from a more general understanding of the cognitive work analysis presented earlier.

This is done because education and training, as in other areas which must consider human factors, it is first necessary to have an understanding of the operator's task and the domain at hand. An analysis of train drivers' operating tasks should have direct implications for the type of education and training needed. They need to be prepared to meet the driving task *as it is represented out in the real situation*. Different training and experience would be needed from what is required today, if for example the signal system were to be presented only through the in cab interface rather than through signals placed at the side of the track.

It is well-known that length and quality of training and experience for safe driving is important. Training and experience, however, cannot be the easy solution for all problems in a complex socio-technical system (Cook & Woods, 1996). Often, persons are related to as the least constrained cause of accidents, which often leads to the disillusioned belief that they are the most avoidable (Hollnagel, 2005b). Traditionally, many industries and enterprises have assumed that it is possible to 'fix' or prevent accidents by providing sufficient training, or by initiating new rules and regulations. More training and detailed regulations have been proven to be an inadequate solution to safety problems, and are not solid long lasting barriers (Hollnagel, 2005b; Health and Safety Executive, 2002). Individuals have a limited capacity when it comes to what they can be 'trained' to do, and what they can remember and use of new regulations or instructions. The specific amount of information that can be dealt with is quite often dependent upon the individuals themselves, but in general we can say that system safety is compromised when barriers such as training, education, and rules and instructions are relied on too heavily in any system.

The current systems analysis suggests that today there are many safety barriers which are dependant upon train drivers' skills, rules, knowledge, and experience. The high degree of context dependency we have observed in the Norwegian rail system demands a high degree of education (e. g. learning the signal and signing system, rules and regulations related to individual enterprises procedures, how to calculate how long it takes to stop the train, etc.) in addition to a great deal of actual hands-on driving experience in specific contexts (hands on experience with the interface, location of signals, weather conditions, what to do when passengers run towards the train because they are late, etc.).

4.2.5.1 Preliminary summary of training and education analysis

It is difficult to conclude whether or not the training and experience drivers have today is sufficient. It may be possible to acquaint drivers with specific situations through, for example, situation-based simulator training in order to increase safety. It may be that drivers, especially inner city drivers, should only drive one specific route due to the high degree of context dependency in the system. However, it may also be that drivers *need* to have a variety of routes to avoid burnout due to the intensity and demands connected with certain routes – something which may be a sign that drivers' capacity to 'learn more' is being exceeded.

Specific scenario-based simulator training should be considered in the future to help cope with the context-dependant rail system which exists today. However, it is quite important to consider carefully *how* simulation should be used. It is suggested that scenario training should be used as a supplement to today's training, and not as a replacement for real-life experience. Nor should simulator training be a kind of 'punishment' for drivers who experience difficulties. If simulation is to be used as a testing situation, the design of the test and evaluation methods should be carefully considered before implementation as there are several challenges related to using simulation in testing or evaluation. It is important that simulators are not used haphazardly. Human factors research on simulator training should be referred to during the design and planning processes. Important aspects that should be clarified before implementation are the task to be simulated, who is to be trained, and how the training should be carried out (Shepherd & Marshall, 2005). In addition, the effects of the use of simulators on other aspects of the MTO system should be considered prospectively – *before* implementing on a large scale.

5 Concluding Remarks

This study has provided an initial systems analysis of the Norwegian railway system. In Part 1, three system critical situations have been identified in the train driving tasks, namely *leaving a station, out on route, approaching a station*. In addition, *switching* was suggested as a fourth driver operation task. Information regarding how drivers use information in these system critical situations may be useful when the human factors are related to various aspects of the MTO system. Future studies may validate whether shifting is a system critical situation in train operation. In the future, it may also be useful to consider train operation in situations other than rural driving routes.

Several system weaknesses and areas for improvement have been discussed, and suggestions for improvement have been put forth. Several suggestions for future research have been made. The systemic view of man-machine interaction in complex socio-technical systems sees the possibility of failure as an attribute of the context. Hollnagel (2005b) finds it useful to think of the relationship between common performance conditions (the system's context) and human error probability in the way one thinks of radio waves, First-generation human reliability analysis methodologies and traditional human factors research based upon human information processing would see human error probability as the 'signal' and the performance shaping factors (the system's context) as 'noise' affecting the 'signal'. However, in second-generation human reliability analysis, such as a systems analysis, the relationship is reversed. Here, the 'signal' is the common performance conditions, while the 'noise' is human error probability. If the system 'signal' is strong and reliable, demands and resources are compatible, and working conditions fall within normal limits, human performance variation should fall within safety boundaries and be more reliable. However, if the system 'signal' is chaotic and unreliable, demands vary, resources may be inadequate, and working conditions may at times be sub-optimal, and the result will be less reliable safety performance. Today, it appears that the Norwegian railway system 'signal' (or performance shaping factors) creates quite an irregular and context-dependant working situation for drivers. To maintain system safety and reduce unwanted events, the current Norwegian railway system is sub-optimal. In such a situation or context, it is not unlikely that human performance variation could result in unwanted events. Drivers may make attempts to relieve the efficiency-thoroughness trade-off's they are presented with in the current situation and safe performance may vary considerably according to the actions individual drivers make.

As this initial system analysis has revealed several weaknesses within the Norwegian rail system which should be improved, there is reason to believe that a more comprehensive systems analysis study would be quite useful in the future. This study has shown that a systemic model of human behaviour can be useful in the Norwegian rail domain, and the use and validation of such models in this industry should be considered in the future. More specifically, modelling train driver performance within the terminology of specific models of human behaviour may be able to explain how the joint cognitive systems in the railway act to achieve their goals, while at the same time responding to events in the environment (see for example the Contextual Control Model (COCOM): Hollnagel, 1983; 1997; 1998; 2002; and Extended Control Model (ECOM): Hollnagel & Woods, 2005). In addition, several methodologies to assist accident prevention have been presented. More specifically, an incident report guide and form for train driving common performance conditions has been developed and presented. Also, the CI is suggested as a practical and helpful tool when collecting information relevant to accident and incident investigation. Further research may serve to further develop and validate the applicability of these methods

References

Borgersen, E. (2001). *Trafikkregler ved jernbanen: en innføring I sikkerhetstjeneste og FO jernbanens "stammespråk".* NSB school internal document

Cook, R. I. & Woods, D. (1996). Adapting to New Technology in the Operating Room. *Human Factors*, 38(4), 593-613.

Dekker, S. (2002). The Field Guide to Human Error Investigations. Ashgate, Aldershot, UK.

Endsley, M. R. (1988). Situation awareness global assessment technique (SAGAT). Proceedings of the National Aerospace and Electronics Conference. 789-95. new York: IEEE.

Endsley, M. & Garland, D. J. (2000). Situation awareness analyis and measurement: analysis and measurement. *New York: LEA*

Endsley, M., Bolte, B. & Jones, D. G. (2003). Designing for situational awareness: An approach to user-centered design. *CRC Press*.

Fisher, R. P., Geiselman, R. E. & Amador, M. (1989). Field test of the cognitive interview: Enhancing the recollection of actual victims and witnesses of crime. *Journal of Applied Psychology*, 74(5), 722-727.

Flach, J. M. (1995). The Ecology of Human-Machine Systems: A Personal History. In Flach, J. M., Vicente, K. J., Hancock, P. A. Caird, J. *Global Perspectives on the Ecology of Human-machine Systems*. Lawerence Erlbaum Associates, 1-13.

FOR 2001-4-12 nr. 1336. Forskrift om signaller og skilt på statens jernbanenett og tilknyttede private spor (signalforskriften).

FOR 2006-12-06 nr 1356: Forskrift om krav til sporvei, tunnelbane og forstadsbane, og sidespor m. m. (kravforskriften)..

Geiselman, R. E., Fisher, R. P., MacKinnon, D. P. & Holland, H. L. (1985). Eyewitness memory enhancement in the police interview: Cognitive retrieval mnemonics versus hypnosis. *Journal of Applied Psychology,* 70, 401-412.

Health and Safety Executive. (2002). Techniques for addressing rule violations in the offshore industries. Offshore Technology Report 2000/096. Available at http://www. hse. gov. uk/RESEARCH/otopdf/2000/oto00096. pdf.

Hollnagel, E. (1983). Why "Human Error" is A Meaningless Concept. Position Paper for NATO Conference on Human Error, August 1983, Bellagio, Italy.

Hollnagel, E. (1997). Cognitive ergonomics: It's all in the mind. *Ergonomcis,* 40(10), 1170-1182.

Hollnagel, E. (1998). Cognitive Reliability and Error Analysis Method (CREAM). Amsterdam: Elviser.

Hollnagel, E. (2002). Cognition as Control: A Pragmatic Approach to the Modelling of Joint Cognitive Systems. Unpublished manuscript, available on http://www. ida. liu. se/~eriho/images/IEEE_SMC_Cognition_as_control. pdf [last checked 30. 03. 2007]

Hollnagel, E. (2004). *Barriers and Accident Prevention.* Aschgate Publishing Limited, Hampshire

Hollnagel, E. (2005a). *CREAM accident analysis: Separating Causes from consequences.* Presentation Presented at CREAM/DREAM workshop. Institute of Transport Economics, Oslo.

Hollnagel, E. (2005b). *Understanding accidents: from simple causes to complex coincidents. Presentation.* Presented at CREAM/DREAM workshop. Institute of Transport Economics, Oslo.

Hollnagel, E. & Woods, D. D. (2005). *Joint cognitive systems: Foundations of cognitive systems engineering.* Boca Raton, FL: CRC Press / Taylor & Francis.

Jansson, A, Olsson, E., & Erlandsson, M. (2006). Bridging the gap between analysis and design: improving existing driver interfaces with tools from the framework ofcognitive work analysis. *Cognition, Technology & Work*, 8, 41-19.

Kecklund, L. & TRAIN project group. Sluttraport från TRAIN-prosjektet. Trafiksäkerhet och informationsmiljö för lokförare. Risker samt förslag till säkerhetshöjande åtgärder.

Kirwan, B. & Ainsworth, L. K. (1992). A guide to task analysis. London: Taylor & Francis.

Lawton, R. & Ward, N. J. (2005). A systems analysis of the Landbroke Grove rail crash. *Accident Analysis and Prevention,* 37, 235-244.

Ljung, M., Furberg, B. & Hollnagel, E. (2004). Handbok För DREAM: Driving Reliability and Error Analysis Method.

LOV 2005-06-17 nr 62: Work Environment Act.

Memon, Amina. (1999). Interviewing Witnesses: "The Cognitive Interview," (p. 343-355) in (Ed.) Memon, A. & Bull, R. *Handbook of the Psychology of Interviewing.* John Wiley & Sons

Petersen, J. (2004). Control situations in supervisory control. *Cognition, Technology & Work*, 6, 266-274.

Railway Technical Pages. (2007). Modern Railway Technology. Available at http://www. railway-technical. com/newglos. html#B. [Last checked 24. 04. 2007].

Rasmussen, J. (1986). *Information processing and human-machine interaction: an approach to cognitive engineering.* North-Holand, New York, NY.

Rasmussen, J. , & Vicente, K. J. (1989). Coping with human errors through system design: implications for ecological interface design. *International Journal of Man-Machine Studies*, 31, 517-534.

Rasmussen, J. (1990). T*he role of error in organising behaviour.* Ergonomics 33(10-11), 1185-1199.

Rasmussen, J. , Nixon, P. , and Warner, F. (1990). Human Error as the Problem of Causality in Analysis of Accidents [and Discussion]. *Philosophical Transactions of the Royal Society of London.* Series B, Biological Sciences, 8, 1241, 449-462.

Rasmussen, J. , Pejtersen, A. M. Goodstein, L. P. (1994). *Cognitive systems engineering.* Wiley, New York, NY.

Reason, J. (1990). *Human Error.* Cambridge University Press, Cambridge.

Shepherd, A. & Marshall, E. (2005) Timliness and task specification in designing for human factors in railway operations. *Applied Ergonomics,* 36, 6, 719-727.

Smiley, A. M. (1990). The Hinton Train Disaster. *Accident Analysis Prevention,* 22, 443-455. In Rasmussen, J. , Nixon, P. , and Warner, F. (1990). Human Error and the Problem of Causality in Analysis of Accidents [and Discussion]. *Philosophical Transactions of the Royal Society of London.* Series B, Biological Sciences, 8, 1241, 449-462.

Van der Flier, H. , Schoonman, W. (1988). Railway signals passed at danger. Situational and personal factors underlying stop signal abuse. *Applied Ergonomics*, 19, 135-131.

Vicente, K. (1999). *Cognitive Work Analysis: Towards Safe, Productive and Healthy Computer-Based Work.* Mahwah, NJ: Lawerence Erlbraum Associates.

Vicente, K. (2004). *The Human Factor.* New York:Routledge. .

Whittingham, R. B. (2004). *The Blame Machine: Why Human Error Causes Accidents.* Elsvier Butterworth-Heinemann, Burlington, MA.

Wilson, J. R. & Norris, B. J. (2006). Human factors in support of a successful railway: a review. *Cognition Technology and Work*, 8, 4-14.

Wilson, J. R. & Norris, B. J. , (eds). (2005). *Applied Ergonomics Special Issue: Rail Human Factors.* Elviser.

Wilson, J. R. & Norris, B. J. Clarke, T. & Mills, A. , (eds). (2005). *Rail Human Factors: Supporting the Integrated Railway.* Hants: Ashgate Publishing Limited.

Wilson, J. R. , Cardiner, L. , Nichols, S. , Norton, L. , Bristol, N. , Clarke, T. & Roberts, S. (2001). On the right track: Systematic implementation of ergonomics in railway network control. *Cognition, Technology & Work*, 3, 238-252.

Woods, D. D. , Cook, R. I. (2002). Nine Steps to Move Forward from Error. *Cognition, Technology & Work*, 4(2), 137-144.

Part 2:

Interface Design

4

Ecological Interaction Properties

Thomas Hoff and Kjell Ivar Øvergård

Traditional approaches to human-machine engineering have mainly been based on cognitive psychology, which views the operator as a rational actor that uses problem-solving strategies to interact with an interface. This understanding has been questioned in recent decades, and new approaches have surfaced. One of these new approaches, termed the Ecological Interaction Properties (EIP), is presented in this paper. This approach describes human-machine interaction as a direct-indirect continuum, which is related to the degree of mental information processing that is necessary for an operator to effectively interact with a system. In a direct interaction no mental processing of information occurs, while an indirect interaction only consists of symbolic problem solving. EIP attempts to describe the ecological properties that affect the directness of a human-machine system, i.e. that affect the operator's need for symbolic information processing.

1 Introduction

To ensure usability and safety within human-machine systems, psychological knowledge of human cognitive and perceptual/motor limitations has been used. The main theoretical approach within the field of human-machine systems has dominantly been traditional cognitive psychology (Hoff, 2004; Rogers, 2004). Traditional cognitive psychology has, in effect, mainly used an analogy of humans as information processing devices similar to complex computers (Reisberg, 2001; Gardner, 1985). Within this approach the human mind is seen as separated from the physical body, and it is this mind that makes sense of—and gives meaning to—the external world. The form of the representations in the environment has not been seen as important within the information processing approach, (Hoff, 2004). This is perhaps because the information processing approach has presupposed that meaning arises in the observer's symbolical processing of the input from the environment, and not (as

Gibson would claim) that meaning is already immanent in the form of the objects constituting the world (Gibson, 1979; Turvey, Shaw, Reed, & Mace, 1981). Traditional cognitive psychology has been the starting point for several approaches to human-machine systems engineering (Card, Moran, & Newell, 1983; Norman, 2002; Rasmussen, 1986), and have had a great impact on how to approach human-machine systems engineering (Rogers, 2004).

In recent decades, following Gibson's (1979) seminal work on the ecological approach to visual perception and Merleau-Ponty's (1962; 1964) phenomenological philosophy, psychologists have begun to question the viability of the cognitive level of analysis and the assumption and implications of the disembodied mind (Hoff, 2002; Lakoff, & Johnson, 1999, Varelas, Thompson, & Rosch, 1991). This has led to a search for new theoretical bases within human-machine systems engineering, and several approaches based on ecological psychology have evolved (Flach, Hancock, Caird, & Vicente, 1995; Hancock, Flach, Caird, & Vicente, 1995, Hoff, 2002). The ecological level of analysis has been reflected in many reports on how to design human-machine systems, particularly the Ecological Interface Design (EID) approach (Burns, Kuo, & Ng, 2003; Burns, & Hajdukiewicz, 2004; Vicente, & Rasmussen, 1992).

One of the primary aspects of the EID approach is that human-machine interaction is seen as composed of three qualitatively different levels of interaction: skill-based behavior, rule-based behavior and knowledge-based behavior (Vicente, & Rasmussen, 1992). Skill-based behavior is dominated by perceptual processing and is based on real-time, multivariate, synchronous coordination of physical moments with a dynamic environment (Rasmussen, Pejtersen, & Goodstein, 1994). Rule-based behavior is behavior where the operator has to use previously established rules to plan the operation of the system ahead. Knowledge-based behavior is based on analytical problem-solving strategies in order to arrive at the correct action in relation to a specific goal. This level depends more upon a symbolic processing of information than the other two levels and, contrary to the other two levels, allows the operator to deal with novelty[1] (Rasmussen et al., 1994).

Another perspective on the levels of activity in human-technology interaction is the direct-indirect continuum (Hoff, Bjørkli, & Øritsland, 2002) which is differenced according to the level of explicit internal problem solving that occurs. It is almost a truism that every type of operator behavior in cognitive systems engineering involves cognition (Salvendy, 1998). Hence it would make sense to categorize cases of human-technology interaction on a continuum related to the degree of internal mental cognition occurring at specific moments in time.

A high-quality interface should, according to EID, support all three levels of interaction (Vicente, & Rasmussen, 1992). Rasmussen and colleagues (Vicente, & Rasmussen, 1992; Rasmussen et al., 1994) have described the overarching principles of how these behavior levels should be supported, but what should explicitly be done is not unambiguously described. An example of this is: "*SBB* [Skill-Based

---

1 See Kaufman's (1980) explication of what modes (semantic, visual or open-ended exploration) of information processing are best suited with regard to the novelty of the problem.

Behavior] – *To support interaction via time-space signals, the operator should be able to act directly on the display, and the structure of the displayed information should be isomorphic to the part-whole structure of movements"* (Vicente, & Rasmussen, 1992, p. 598, italics in original).

2   The Ecological Interaction Properties

The Ecological Interaction Properties is a recently developed taxonomy that presents ecological properties related to the direct-indirect continuum of the operator-machine system (Hoff, 2002; Hoff, et al., 2002). The direct-indirect continuum refers to the degree of directness in human-machine interaction. An interaction property is defined as "the smallest element to which a change in that element introduces a change in user experience" (Hoff et al., 2002). The degree of directness can be related to the degree the operator uses automated or skill-based behavior in the operation of the interface. On the direct side of the continuum is an interaction that is fully skill-based, while on the indirect side is interaction that has stalled because the operator must employ problem solving to continue the operation of the system.

In a perfectly *direct* interaction the operator interacts with the tool (or the interface) in a way that fosters a close perception-action coupling, where the interaction continues unhindered by mental workload or the need for problem solving on the operator's part. Tools and interfaces can be incorporated into the operator's bodily and perceptual world (Gibson, 1979; Smitsman, 1995), and they can extend a human's action capabilities (Hirose, 2002) and perceptual capabilities (Hancock, & Chignell, 1995). The boundary of the body can then extend beyond the skin, such as when a blind man perceives the ground through a cane (Bateson, 1972; Hirose, 2002). In a human-machine system that supports direct interaction, the operator extends the functions of his/her physical and mental characteristics into the machine system and lives through the system, rather than *interacting* with the interface, as such.

An *indirect* interaction is one where the operator has to use explicit reasoning and problem solving to effectively interact with the interface/system. The need for problem solving will slow down the interaction since the operator will have to mentally represent significant aspects of the machine environment to be able to proceed with the operation or monitoring of the system (Vicente, & Rasmussen, 1992). Machine-operation based on problem solving is also more error-prone than skill-based operation (Reason, 1990). Hence problem solving is not an optimal level of interaction and should be avoided if more skill-based levels of control can be employed. This said, problem solving is also important in the operation of machine systems, since problem solving allows the operator to deal with novel situations (Rasmussen et al., 1994).

The direct-indirect continuum includes the experiential aspect of human-machine interaction, and as such can function as an evaluative tool for future or current interfaces (Hoff et al., 2002). The EIP consists of dimensions that are both supposed to affect the operator's ability to operate the interface on a skill-based level, and also of dimensions that are purely evaluating tools, which are meant to be used to evaluate the interface in situ, in conjunction with an operator.

2.1 The Philosophical and Theoretical Basis of EIP

The philosophical basis for EIP is based upon ecological psychology (Gibson, 1979) and second-generation cognitive psychology (Lakoff, & Johnson, 1999; Varelas, et al., 1991). It also has close connections to phenomenological psychology (Merleau-Ponty, 1962; 1964) and activity theory (Nardi, 1996).

Within second-generation cognitive psychology, the human mind is seen as an embodied aspect of the individual's reciprocal interaction with its environment. Cognition depends upon the experiences a person has through the sensorimotor capacities of the body, and this experience is itself embedded in a biological, cultural and psychological context (Varelas et al., 1991). The concept *embodied* means that everything we experience is experienced through the body, and because of this there can be no clear distinction between the outer and the inner of human experience (Merleau-Ponty, 1962; Lakoff, & Johnson, 1999). With this starting point, the theoretical basis of EIP transcends the subjective and objective aspects of lived experience. The experiences are *for* a person, *in* a world, and the mind-body dualism introduced by Plato and Descartes is rejected on the basis that the experiencing subject never lives in a subjective *or* objective world. The concepts of subjective/objective taken from logical positivism do not relate to what actually is human experience, as the individual always experiences something for him/her (Giorgi, 1970; Varelas et al., 1991).

The ecological approach claims that researchers should study the interaction between humans and the environment on the ecological level, and not on the level of mental representations (Gibson, 1979). Furthering this argument, researchers have claimed that this also applies to human-machine systems (Hoff, 2002). An ecological analysis starts out with the world as perceived by an observer (Gibson, 1979). This perspective differs to a great extent from the traditional cognitive perspective where information is viewed as objective and that it systematically reduces uncertainty in relation to some subject matter. Rather, it is claimed that there exist some features of the environment that are more salient and meaningful to human observers. This has occurred because humans have co-evolved with a particular environment, and because of this they have become more sensitive to certain aspects of this environment (Hoff, et al., 2002; Gibson, 1979; Lakoff, & Johnson, 1999). Due to this assumption, ecological psychologists may say that the way information is represented in the interface affects how easily an operator can interact with the interface (Smets, 1995).

EIP has been presented as an attempt to investigate how information or meaning should be represented in an interface to induce direct perception.[2] It is important to

---

[2] With the correct use of perception-scaled features in an interface, these features are not represented since they lead to a direct and unmediated pick-up of the meanings inherent in the interface-operator system.

state that the EIP contains *principles for interface design* rather than an actual way of representing information in an interface. The EIP itself is without content; rather, it states the dimensions that could be manipulated to alter user experience and the directness of a human-machine system. It is meant to be a general framework that can be applied to any interface regardless of its employment.

2.2 Explication and Redefinition of the EIP

The EIP consists of eight properties,[3] each relating to a different aspect of the ecology of human-machine systems. These properties are overlapping and are meant to function as both a qualitative evaluating tool to map the directness of a given interface, and as guidelines for how to design new interfaces (Hoff et al., 2002). They include the phenomenological aspects of human-machine interaction and, hence, are related to the user-experience of the given technology.

The EIP can be separated both into properties that can be used to affect the propensity for an interface to induce direct interaction, and into properties that are of a more evaluative nature. Whether the EIP can be used to design interfaces that give altered or better user experiences must be explicated through future research, although theoretical claims show that these properties can embody altered user experiences (Hoff et al., 2002).

2.3 Properties Affecting the Directness of an Interface

These properties are related to aspects of the interface that hypothetically affect the user experience, and the tendency for an interface to induce direct interaction in the human-machine system. These properties are also operationalized in a way that makes them suitable for rigorous experimental testing. They do not involve a specific content, but rather present dimensions that can be manipulated to alter the degree of directness in the human-machine system.

2.3.1 Continuous-Discontinuous Property

This property relates to the representations of information presented in the interface, and to the degree that these representations are *continuous* or *discontinuous*.[4] This is related to whether the information is presented as continuous events that simulate the spatial and temporal aspects of real-life events or whether they are presented in a bit-by-bit fashion (Hoff et al., 2002). It is assumed that normal supervision of the function of a given system probably will benefit from a dynamic continuous display

---

3 The aspects that EIP consist of have not yet been empirically tested, so there is no empirical or statistical information on whether these aspects are dimensions, categories or properties. For this reason the word 'properties' will be used as a label for all EIPs.
4 This continuum has been named Analogue-Digital by Hoff and colleagues (2002), though this name has led to difficulties as such names have cultural connotations related to the form of the equipment rather than to the nature of the representations in the system.

of information (Hoff, & Hauser, in press). This property may be described as a continuum that at one end includes an interface that has continuous information representations, while at the other end it contains a static representation of the state of the system that changes like a photographic slide show.

A display with continuous information representation will probably simulate the natural world in a better way than a discontinuous display, and will as such have a higher ecological validity and foster a more direct user experience. An example of an interface with continuous representation is an analogue clock, where the arms are in continuous motion. This clock allows the user to quickly determine the time (given that s/he has learned the cultural standard of time representation). It also gives a better impression of how much time has passed since the last time the user checked the clock, probably because the arms are laid out in a two-dimensional array which is easy to compare from one time to another ("the short arm has moved from here to there"). This does not apply to the digital clock as it is always in the same spatial placement; the only thing that changes is the numbers that arbitrarily map the current time (Norman, 2002). A digital clock informs the user about the exact time (hours, minutes, and seconds) but lacks the continuous movement and spatial placement of arms. The continuous representation is supposedly superior to the digital one if the task of interest is to know the time in hours and minutes, while the digital one is better when the exact time is required to be known (as in a 100-meter sprint final in the Olympic games).

2.3.2 Specific-Generic Property

This property relates to how many functions a given control has. The relation between the number of functions and the number of controls leads to two types of complexity. With many controls, each of which has one function, the *apparent complexity* (or perceived usability) of the interface will increase (or decrease, in terms of usability). When there are few controls, each with a greater number of possible functions, the *complexity of use* (or actual usability) will increase, since the increased complexity of the controls must be mirrored by a greater complexity of use by the operator (Norman, 2002).

Specific controls will control only one function or one operation domain. An example of a specific control can be the on/off button on a computer. This control has two functions: the first is to turn on the computer when it is inactive, while the other is to turn the computer off when it is operational. Both of these functions are related to the same work-domain: they regulate the operative level of the computer. The control does not have any other functions, and the need for mental mapping of its functions is not necessary. An example of a control with a generic functioning is the iDrive Controller™ manufactured by Immersion [5] and employed by BMW [6]. The

---

5 Immersion Corporation (2003). BMW case study. [27.10.03]. http://www.immersion.com/automotive/applications/bmw_case_study.php
6 The BMW 7 Series (2003). Hundreds of Controls in One: The iDrive Controller. [27.10.03]. http://www.bmw.com/e65/id14/3_a91_idrive.jsp

iDrive controls several hundred functions, among which are the car stereo system, telephone, DVD, the in-car television, a speech-controlled navigation system, and an on-board computer system [7].

It is presumed that a specific control has more ecological validity than a generic control, and that the interaction will be more direct when specific controls are employed rather than using one generic control. The use of a generic control removes some of the complexity in the interface, but a reduced complexity in the interface must often be adjusted with a higher complexity in the actions of the user. Often, an interface has several possible functions, but if the interface has few controls (which then have several functions each) the complexity of the use of the control will increase. The increased complexity that comes with more complex controls leads to the use of problem-solving strategies where the operator must actively search for knowledge before an effective interaction can pursue. An example of a device with complex controls is a mobile phone (e.g. Siemens C55) with several functions, such as Internet access, short message service (SMS), calculator, and alarm function. In such a mobile phone the controls that are used to key in the phone numbers must also be used when writing an SMS, when setting the alarm, and when navigating the menu system.

2.3.3 Perceptual/Motor Scaling Property

This property relates to how the design of the interface is fitted to the human sensorimotor capabilities. It can be divided into two different aspects. These aspects are closely related to the use of anthropometric data and human factors guidelines, but are different in that they focus on the ecology of the total experience of interaction, which includes bodily movement and different sensory modalities (Hoff et al., 2002). On the other hand, they differ from anthropometric data because they are not only anatomical but also functional. This is because the size of many environmental objects must be fitted not only to the size of the operator's body, but also to the task and the actions of the operator (Warren, 1995).

The first aspect is how the interface is scaled in regard to the limitations of the human sensory system, e.g. whether the interface contains (or displays) light (lux and color contrast), sound (Hz, pitch, timbre, etc.), tactile stimulation (surface texture, etc.) and haptics (when manipulating controls, etc.) that are within the *'perceptualrange'* of the human operator.[8] An interface that induces direct interaction should be fitted to the biological limitations of the human sensory system, e.g. it should be fitted to the perceptualrange (Hancock, & Chignell, 1995).[9]

---

7   The BMW 7 Series (2003). Intelligent Solutions: The BMW iDrive Comfort Zone. [27.10.03]. http://www.bmw.com/e65/id14/3_a9_idrive.jsp
8   Olfactory and gustatory feedback is not mentioned here, as these sensory modalities are not equally important to the other senses when interacting with computer systems.
9   The perceptualrange is the part of the environment that humans can perceive without the help of tools or technology, e.g. it is the range for unaided perception (Hancock & Chignell, 1995).

The other aspect is how the interface and its controls are scaled to the human body and the movement pattern of the operator. This relates to the size, shape, weight, and surface structure of the interface and its controls. Research has been carried out on the ecology of human movement and action in an environment. Body-scaled biomechanical invariants have been found in stair climbing (Warren, 1984), when passing through apertures (Warren, & Whang, 1987), in sitting height (Mark, & Vogele, 1987), and in grip transitions (Cesari & Newell, 2000).

These invariants can be said to afford[10] specific types of actions from the observer (Gibson, 1979). By locating these body-scaled invariants that guide (and can be used to predict) human behavior, researchers can produce more ecologically correct surroundings (Warren, 1984). If it is correct to assume that such affordances exist in the natural world as perceived by humans, and that these guide our actions, it should not be too bold to assume that the use of these biomechanical invariants can also improve interface design.

2.3.4 Perceptual Richness Property.

This property relates to the number of senses that are involved in the interaction with the interface (Hoff et al., 2002). The human body, through co-evolution with the environment, is adapted to multimodal sensory input. The five perceptual systems (orienting, auditory, haptic, taste and smell, and visual) are interrelated, and they work as a system where they all contribute to the perception of an event or an object (Gibson, 1966). A multimodal and perceptually redundant event will involve several senses and as such also includes several types of stimulus information that can be picked up by the observer (Gibson, 1966). If an event is perceptually 'poor' and only consists of one sensory modality, less information is available for the observer to pick up. This may lead to a need for additional stimulus information as the information from the event is too scarce and/or impoverished to be the basis for effective perception or recognition.

A computer interface often gives feedback in only one sensory modality, which can lead to problems as the operator may have to use cognitive inference (problem solving or rule-use) to effectively understand the information presented by the interface.

2.3.5 Motoric-Sensoric Mapping Property.

This level relates to the degree that the representations in a system are spatially related to the actions of the operator. The actions of the operator occur in a four-dimensional space[11] where objects generally conform to the size, speed and direction of

---

10 Affordances (a neologism first presented by Gibson, 1979) describe what can be done with a given object. They define the actions made possible for a given observer with a given object (Turvey et al., 1981). One can also say that an affordance specifies the action capabilities of an object, and can only be found on an ecological level (Gibson, 1979, see also Zaff, 1995).
11 The four dimensions are height, width, depth, and time. Four-dimensional space is used to denote that the actions of a human are always spatially and temporally placed.

these movements. Consider the example of moving a book, where the book held in the hand will conform in a one-to-one relation with the arm/hand's movements. It is not only in the physical world that there is conformity between the movement of the observer and changes in the world, it also seems to be the case with visual imagery, where researchers have found that the time taken to scan a visual image is correlated with the time taken to scan the actual object. It seems that visual images conserve the functional size of the imagined object (Kosslyn, Ball, & Reiser, 1978). From this it might be argued that both the physical and mental movements of a person both reflect and are based upon the relation between the world and the perceiving individual (see also Lakoff, & Johnson, 1999, for a similar argument).

In a computerized or virtual system, the movements of the observer may not lead to the same movements of the representations in the interface, or of the parts of the machine. This difference can be in regard to direction (as with a computer mouse) or in relation to the size of the movements (in virtual environments). By making the representations in the machine-system as close as possible to how the motoric-sensoric mapping are in the natural environment, it is presupposed that breaks in perception-action cycles can be minimized. A better simulation of the natural occurrences of motor-sensory mapping may be important for increasing the efficiency and usability of interfaces. The hypothesis that discrepancies between an observer's movements and the representations in the system may lead to a break in the perception-action cycles should be tested in future research.

2.4 Evaluative Properties of Interfaces

The following EIPs do not state which aspects one could alter to affect the user experience, but rather involve more holistic aspects of the total operator-interface ecology. It is presupposed that the properties presented are vital to induce a direct interaction between the operator and the system.

2.4.1 Input/Output Property

This property relates to the degree that the operator perceives a gap between the input given and the output received. This gap could either be a temporal or a qualitative difference between the input and the output. A temporal lag could be exemplified by an overloaded computer, which does not have the CPU capacity to handle all the information at once, something that leads to a delay in responses to the operator's commands (input). A qualitative difference could be that the output consists of sensory modalities that are not prospected by the operator, given the modalities of the input. Other qualitative differences could be a plane-mapping discrepancy as presented by the use of a mouse to move the cursor on a computer screen (Hoff et al., 2002).

This property is not directly related to any specific factors that affect the directness of the interface, since it does not state any given dimensions to manipulate. Rather, it is a dimension that is to be explored in situ with an operator actually operating the interface. It is presupposed that a direct interaction would be more likely when there are no perceived gaps between input and output. However, with a difference between input and output, a greater need for mental processing (use of working memory to keep track of recently executed actions) is expected, something that leads to a less direct interaction in human-machine systems.

2.4.2 Physical-Inferential Property

This property is composed of three different aspects relating to the representations in the interface and the operator's actions and the consequences of these actions (or the representations of these consequences). These properties overlap to some degree and they are possibly not even exhaustive, but they will be described separately to better explain the important aspects of this ecological interface continuum.

The *first* aspect is *prospection of consequence*. This is the semantic relation between the controls and their functions. This aspect relates to whether the consequences of the use of the control are mediated by the layout of the control. The first aspect can be described by the question "In what way do the control's layout inform the user about the consequences of manipulating the control?" A direct interface in this respect is an interface where the outcome of manipulation is perceived without any cognitive inference. An indirect interface is one where the operator has to employ systematic problem solving before s/he can use the control effectively. The concept of prospective control (Hofsted, 1996; Lee, 1993) becomes important here, as the design of the interface can help the operator plan ahead when operating the system. As Hofsted (1996, p. 33, italics added) claims, "All actions are geared to the future and *controlling them requires knowledge of upcoming events*. We continuously need to know what is going to happen, both within the near and the far future, in order to plan our activities and to coordinate our movements".

The *second* aspect is *representation of consequences*. This relates to whether there is a *meaningful relation* between the *output* and the *representations of the output*. This aspect can be described by the question "Do the representations of the results of the machine's actions (given that the operation does not lead to physical results) relate in a meaningful way to the actual result?" A direct interface is one where the operator will know which operations the system has performed from viewing the representations of the results. These representations could be, for example, graphs, symbols or pictograms.

This problem is highly related to human-computer interaction, as the output in a computerized system is often of a digital, graphical nature. Physical, real-life devices often have direct output. A hammer, for example, has a direct physical output that includes several sensory modalities (haptic, visual and auditive senses), that stand in a natural relationship to the input (hitting with the hammer), whereas the consequences in a digital computerized system often have to be inferred because they are "hidden inside" the system, and the feedback representing the consequence often is in an abstract/symbolic relationship with the consequence. It is also possible that the system does not give any feedback to the operator.

The *third* aspect is *response mode guidance*. It relates to the degree that the control informs the operator about what type of response is needed to effectively manipulate the control: whether the control should be pushed or turned, pulled or pressed, and which part of the body it is designed for. This aspect is close to what Norman (2002) call the "gap of execution", where the operator can perceive what actions are possible and what actions should be performed. This could be an example of what is normally called an affordance (Gibson, 1979). In a computerized system, many symbols or icons are arbitrary and do not afford the operator information as to how to manipulate them in a successful way. By simulating the natural affordances in a computer system, designers may improve the usability of these systems (Norman, 2002).

A direct interface will support prospective control and planning ahead better than an indirect interface, as the direct interface will contain affordances that inform the user about the correct actions to take. Designing interfaces that contain affordances or artifacts that help the operator to plan ahead, and that support the operator in a way that s/he knows what the consequences of his/her actions will be, is of vital importance. An interface that minimizes the cognitive load on the operator is presupposed to induce more direct interaction. With increasing semantic discrepancies between the controls and their functions and the output and the representations of the output, it could be expected that the operator would have to employ cognitive inference to be able to use the interface.

Hence, it is important that this ecological property is explored in situ, in a representative environment(s) and with a representative operator (see also Faulkner, 2000).

2.4.3 Syntax-Semantic Property

This property describes the relation between the structure of the interface and the total functions which the interface represents. *Syntax* refers to the structure of the interface and *semantics* refer to the total functioning of the system. A direct interface is one where the structure of the interface informs the operator of the functions of the system without any cognitive inference (Hoff et al., 2002).

This ecological property is closely related to the physical/interferential property, but it is of a slightly higher level. This is because the syntax-semantic property relates to how the total interface affords the functions of the system, while the physical-inferential property refers to how a single control can afford the action of the operator, the consequences of its use, and how the consequences are represented in the interface. It is possible that this property could be incorporated in the physical-inferential property, but then as an overarching property that relates more to the purpose of the interface/system than to the functions of the given controls. Further theoretical and empirical work will hopefully solve this challenge.

2.5 Discussion

The current presentation of EIP is an extension of the work of Hoff (2002). EIP is a rather new approach, which attempts to cover uncharted theoretical and practical ground. It is therefore evident that the next step in the development is to test the

EIP empirically. Some of the properties (continuous-discontinuous, specific-generic, perceptual richness, perceptual/motor scaling, and motoric-sensoric scaling) should be tested experimentally to see whether they actually affect the directness of human-machine interaction. These properties are described in a way that allows for easy operationalization and later testing. The remaining properties (input-output, physical-inferential and syntax-semantic) have also been altered, but are still qualitative and as such would demand qualitative phenomenological analyses to be tested, as they are described here. Further theoretical development of these properties should be related to—and dependent on—empirical results from both experimental and real-world studies.

Other aspects in the development of the EIP is whether they can be used as a practical tool to design interfaces that are effective and usable, and whether they can be used as evaluative tools for existing technological design. Such a practical approach should be combined with an empirical approach. These approaches to the EIP and interface design should be viewed as complementary and a necessary part of the development of an empirically supported practical tool for interface design.

3 Conclusions

The Ecological Interaction Properties are still under theoretical and practical development and their actual effect on the design of human-machine systems is still to be explored. Since the EIP is mainly related to what Vicente and Rasmussen (1992) have called the skill-based behavior level, it must be said that knowledge about the EIP is necessary, but not sufficient, to design human-machine systems. Accordingly, the EIP should be used within a design methodology, e.g. such as Vicente's (1999) Cognitive Work Analysis, to cover all areas of usability and safety problems found in complex socio-technical systems. Areas which, at the present, are not covered by the EIP, are semantic differentiation of controls, spatial and semantic layout of controls, how to support effective inter-operator communication, and other properties related to Vicente and Rasmussen's (1992) knowledge-based behavior level.

It is apparent that the EIP can fulfill two different purposes. The first is that the properties can be used to guide the design of new interfaces that induce fluent, skill-based interaction. The second purpose is that the properties can be used to evaluate existing interfaces. Hopefully, the EIP will be able to show what aspects of an interface lead to a breakdown in the operator's perception-action cycles. Further research will reveal how the EIP affects usability and safety in human-machine systems.

References

Bateson, G. (1972). *Steps to an Ecology of Mind: Collected Essays in Anthropology, Psychiatry, Evolution and Epistemology*. London, UK: Chandler Publishing Company.

Burns, C. M., & Hajdukiewicz, J. (2004). *Ecological Interface Design*. CRC Press.

Burns, C. M., Kuo, J. & Ng, S. (2003). Ecological interface design: a new approach for visualizing network management. *Computer Networks*, 43, 369-388

Card, S. K., Moran, T. P., & Newell, A. (1983). *The Psychology of human-computer interaction*. Hillsdale, NJ: Lawrence Erlbaum Associates.

Cesari, P. & Newell, K. M. (2000). Body-Scaled Transitions in Human Grip Configurations. *Journal of Experimental Psychology: Human Perception and Performance*, 26, 1657-1668

Faulkner, X. (2000). *Usability Engineering*. London, UK: Macmillan Press Ltd.

Flach, J., Hancock, P. A., Caird, J., & Vicente, K. J. (1995). *Global Perspectives on the Ecology of Human-Machine Systems*. Hillsdale, NJ: Lawrence Erlbaum Associates.

Gardner, (1985). *The Mind's New Science – A History of the Cognitive Revolution*. Basic Books.

Gibson, J. J. (1966). *The Senses Considered as Perceptual Systems*. Boston: Hougthon Mifflin Company.

Gibson, J. J. (1979). *The Ecological Approach to Visual Perception*. London, UK; Hougthon Mifflin Company.

Giorgi, A. (1970). *Psychology as a Human Science: A Phenomenologically Based Approach*. New York; Harper & Row, Publishers.

Hancock, P. A. & Chignell, M. H. (1995). On Human Factors. In J. Flach, P. Hancock, J. Caird, and K. Vicente (Eds.) *Global Perspectives on the Ecology of Human-Machine Systems*. Hillsdale, NJ: Lawrence Erlbaum Associates.

Hancock, P. A., Flach, J., Caird, J., & Vicente, K. J. (1995). *Local Application of the Ecological Approach to Human-Machine Systems*. Hillsdale, NJ: Lawrence Erlbaum Associates.

Hirose, N. (2002). An ecological approach to embodiment and cognition. *Cognitive Systems Research*, 4, 289-299.

Hoff, T. (2002). *Mind Design: Steps to an Ecology of Human-Machine Systems*. Dr. Polit. Dissertation, Department of Psychology and Department of Technological Design Engineering, Norwegian University of Science and Technology, Trondheim, 2002.

Hoff, T. (2004). Comments on the Ecology of Representations in Computerised Systems. *Theoretical Issues in Ergonomics Science*, 5(5), 453-472.

Hoff, T. & Hauser, A. (in press). Applying the Ecological Approach to Interface design on Energy Management Systems: Developing a Compact System State Display (CSSD). *PsychNology*.

Hoff, T., Øritsland, T. A. & Bjørkli, C. A. (2002). Ecological Interaction Properties. In P. Boelskifte, J. B. Sigurjonsson (Eds.). *Proceedings of NordDesign 2002*, (pp. 137-151).

Hofsted, C. (1996). Development of prospective control as the basis of action development, *Infant Behavior and Development*, 19, Supplement 1, 32.

Jamieson, G. A. & Vicente, K. J. (2001). Ecological interface design for petrochemical applications: supporting operator adaptation, continuous learning, and distributed, collaborative word. *Computers and Chemical Engineering*, 25, 1055-1074.

Kaufmann, G. (1980). *Imagery, language and cognition: toward a theory of symbolic activity in human problem-solving*. Bergen, Norway: Universitetsforlaget.

Kosslyn, S. M., Ball, T. M. & Reiser, B. J. (1978). Visual images preserve metric spatial information: Evidence from studies of image scanning. *Journal of Experimental Psychology: Human Perception and Performance*, 4, 47-60.

Lakoff, G. & Johnson, M. (1999). *Philosophy in the Flesh: The Embodied Mind and its Challenge to Western Thought*. New York, NY: Basic Books.

Lee, D. N. (1993). Body-environment coupling. In U. Neisser (Ed.), *The perceived self: ecological and interpersonal sources of self-knowledge* (pp. 43–67). Cambridge: Cambridge University Press.

Mark, L. S. & Vogele, D. (1987). A biodynamic basis for perceived categories of action: A study of sitting and stair climbing. *Journal of Motor Behavior*, 19, 367-384.

Merleau-Ponty, M. (1962). *Phenomenology of Perception.* (C. Smith trans.). London; England: Routledge & Kegan Paul. (Original work published 1945).
Merleau-Ponty, M. (1964). *Le visible et l'invisible.* Paris: Gallimard
Nardi, B. A. (Ed.) (1996). *Context and Consciousness: Activity Theory and Human-Computer Interaction.* Cambridge, Massachusetts: The MIT Press.
Norman, D. A. (2002). *The Design of Everyday Things.* New York: Basic Books.
Rasmussen, J. (1986). *Information processing and human-machine interaction: an approach to cognitive engineering.* New York: North Holland.
Rasmussen, J., Pejtersen, A. M. & Goodstein, L. P. (1994). *Cognitive Systems Engineering.* New York: John Wiley & Sons, Inc.
Reason, J. (1990). *Human Error.* Cambridge, UK: Cambridge University Press.
Reisberg, D. (2001). *Cognition: Exploring the science of the mind*, 2nd ed. New York : Norton
Rogers, Y. (2004). New Theoretical Approaches to HCI. *Annual Review of Information Science and Technology,* 38(1), 87-143.
Salvendy, G. (1998). A response to John Dowell and John Long, 'Conceptions of the cognitive engineering design problem'. *Ergonomics,* 41(2), 140-142.
Smets, G. J. F. (1995). Industrial Design Engineering and the Theory of Direct Perception and Action. *Ecological Psychology,* 4, 329-374.
Smitsman, Ad. W. (1995). Affordances and the Practice of Industrial Design Engineering: Comments on Smets's Presentation. *Ecological Psychology,* 7, 375-378.
Turvey, M. T., Shaw, R. E., Reed, E. S. & Mace, W. M. (1981). Ecological laws of perceiving and acting: In reply to Fodor and Pylyshyn (1981). *Cognition,* 9, 237-304.
Turvey, M. T., Shockley, K. & Carello, C. (1999). Affordance, proper function, and the physical basis for perceived heaviness. *Cognition,* 73, B17-B26.
Varelas, F. J., Thompson, E., & Rosch, E. (1991). *The Embodied Mind: Cognitive Science and Human Experience.* Cambridge; Massachusetts: The MIT Press.
Vicente, K. J. & Rasmussen, J. (1992). Ecological Interface Design: Theoretical Foundations. *IEEE transactions on Systems, Man and Cybernetics,* SMC-22, 589-606.
Warren, W. H. Jr. (1984). Perceiving affordances: Visual guidance of stair climbing. *Journal of Experimental Psychology: Human Perception and Performance,* 10, 683-703.
Warren, W. H. Jr. (1995). Constructing an Econiche. In J. Flach, P. Hancock, J. Caird & K. Vicente, (Eds.), *Global Perspectives on the Ecology of Human-Machine Systems.* (pp. 210-237), Hillsdale, NJ: Lawrence Erlbaum Associates.
Warren, W. H. Jr. & Whang, S. (1987). Visual guidance of walking through apertures: Body-Scaled information for affordances. *Journal of Experimental Psychology: Human Perception and Performance,* 13, 371-383.
Zaff, B. S. (1995). Designing with Affordances in Mind. In J. Flach, P. Hancock, J. Caird & K. Vicente, (Eds.), *Global Perspectives on the Ecology of Human-Machine Systems.* (pp. 238-272), Hillsdale, NJ: Lawrence Erlbaum Associates.

5

On Representations of In-Vehicle Information Systems – Effects of Graphical (GUI) versus Speech-Based User Interfaces

Cato A. Bjørkli, Thomas Hoff, Gunnar D. Jenssen,
Trond A. Øritsland and Pål Ulleberg

The human cognitive system places constraints on representations of human-machine systems. This simulator-based study compares the effects of a traditional graphical interface (GUI) and a speech-based interface of an in-vehicle information system, in terms of perceived subjective workload as measured by NASA-TLX, as well as driver performance as measured by speed and lateral positioning. Both interfaces proved to place significant burdens on the driver compared to perceived workload of normal driving, as well as in relation to the experimental baseline. Despite this, in terms of workload, the hand-held interface proved to be poorer on several dimensions than the speech-based interface. The paper considers theoretical, experimental and practical implications of interface design in human-machine systems.

Keywords: cognitive engineering, interface design, in-vehicle information systems, representations, traffic psychology

1 Introduction

The trend towards integration of information technology (IT) into vehicles represents a significant contribution to the in-vehicle environment, as it offer drivers the possibility of dynamic communication of information. Obvious examples include the possibility for drivers to receive specific information about choice of route, road conditions, accidents, and sights. Such systems are particularly useful within the transportation sector, in that commercial drivers have the possibility of staying in close contact with their firm and the end-line customers during a given assignment. The possibilities put forward by IT come at a cost, however, because any activity other than those directly related to the operation of the vehicle potentially impacts on safety.

The operation of IT devices adds tasks that are comprehensive and time consuming compared to discrete operations regarding driving as such. This is traditionally believed to increase the operator's workload and further affect driving performance (Lai et al., 2001).

In order to prevent IT from becoming a major safety threat, during the recent years international standards and national regulations for the design of such systems have been developed. One of the most extensive attempts to formalise human-factors demands of such systems is the british Transport Research Laboratory's safety checklist for the assessment of in-vehicle information systems (BSi, 1996). It is clear, however, that more than human-factors and guidelines are needed in order to develop optimal user interfaces for in-vehicle information systems.

Human-factors guidelines are simplified heuristics based on knowledge of basic human cognitive processes. They are useful and necessary in coordinating the surface look and feel of user interfaces. However, when the aim is to design entire information systems, these guidelines no longer provide sufficient knowledge. Here, contemporary knowledge of the functioning of the human mind and its relation to its external context must necessarily come into play. Following Batesons' (1972) 'blind man's stick' argument, technology should be an extension of fundamental human capabilities. Technology aids humans to perform beyond limitations of their biology and psychology. In this view, technology redefines the boundaries of human capabilities of perception and action. The further these boundaries are extended, the more complex are the systems needed to support efficient use, which further increases the intricacy of the representations of control interfaces (Hancock and Chignell, 1995). The matter of enabling humans to perform beyond their basic capabilities – from unaided to aided performance – presents design challenges that human-factors guidelines are not sufficiently equipped to meet.

1.1 Theoretical perspectives

Within transportation research, driving behaviour is often understood as a hierarchically organised activity (Michon, 1989; Wickens, 1991; Keskinen, 1996). The organisation of behaviour generally refers to three levels: the highest level refers to actions concerning strategic decisions. The middle level is a tactical level including responses to the immediate traffic scenario, whereas the lowest level is an operational level consisting of specific acts such as shifting gears or steering the vehicle. This description is in line with the skills, rules and knowledge (SRK) taxonomy originating from cognitive engineering research (Rasmussen, 1983; Vicente and Rasmussen, 1992). The basic concepts in the taxonomy are: skill-based behaviour, equivalent to the operational level; rule-based behaviour, equivalent to the tactical level; and knowledge-based behaviour, equivalent to the strategic level. The behaviours refer to a type of cognitive control, and the taxonomy can be grouped in two general categories: on the one hand, there is fast, perceptual processing, and on the other hand there is slow, analytical problem solving. Perceptual processing is reckoned to be effortless, whereas analytical problem solving is more laborious and comprehensive. Translated to the traditional workload terminology, one can expect perceptual processing, as in skill-based behaviour, to place less strain upon the operator than the analytical problem solving of knowledge-based behaviour. These two cognitive control modes may be thought of as layers in a hierarchical organisation. Behaviour constantly fluctuates between the states of control as the driver interacts with his/her context, where each state functions as a qualitatively different way of processing information.

As pointed out by Haigney et al. (2000), driver performance is rooted in the relationship between cognitive resources, the specific demands of driving, and any secondary tasks imposed on the driver. The capacities of the driver (e.g. attention resources, driver experience, memory) are the foundation of this close relationship. Yet, the outcome of driver performance is not exclusively dictated by driver states. The driver relates to the task in the sense that task characteristics activate different levels of cognitive control as well as motor behaviour. Further, how the information of a task presents itself to the driver will determine what type of control and behaviour is activated. The interface conveys the task and may present itself as signals, signs or symbols (Vicente and Rasmussen, 1992). The interplay between the interface and the driver behaviour thus becomes a main area of interest. Due to its presentation of information, an interface will influence the amount and type of mental effort, and thereby set the boundaries for how much workload is imposed on the driver. An important point in this line of argument is that the relationship between representations and cognitive control mode is not fixed. Representations may engage several cognitive control modes and do not exclude any of them. For example, the speedometer represents the current vehicle speed. The skilled driver can in one situation just take a short glimpse at the instrument to get an idea of the current speed, while in another situation clearly reading the exact speed indicated by the needle. The former use of the instrument refers to an attuned and perceptual way for relating to the interface similar to a low-level cognitive control mode, whereas the latter refers to an analytical, information processing activity. This underlines the fact that interfaces may afford all three cognitive control modes. Yet, there is a fundamental ecology of representations in the sense that they favour and support some modes rather than others (Hoff, 2004). The design of interfaces needs explicitly to incorporate this ecology, in order to support safe and efficient in-vehicle information systems.

1.2 Empirical background

Prior research has indicated a number of associations between the use of mobile phones, workload and driver performance. On a general basis, epidemiological studies have shown a clear association between the use of in-vehicle information systems and an increased frequency of accidents (Violanti and Marshall, 1996; Redelmeier and Tibshirani, 1997; Stevens and Minton, 2001). Drivers who use a mobile phone while driving are generally four to five times more likely to have an accident than drivers who do not use a mobile phone.

A number of studies have investigated specific aspects of how in-vehicle information systems affect the driver and driving performance. Several studies have indicated a relationship between subjective experience of workload and the use of mobile phones while driving. Self-reports, such as the NASA-TLX (Hart and Staveland, 1988), have indicated an increased level of subjective workload when using a mobile phone while driving (Fairclough et al., 1991; McKnight and McKnight, 1993; Alm and Nilsson, 1995). This applies both to the traditional hand-held mobile phones as well as hands-free mobile phones.

There are strong indications that an increased workload will manifest itself in terms of various changes in driver performance. The primary dimensions of performance are the lateral and longitudinal movements of the vehicle, expressed in terms of mean and standard deviation of both speed and lateral positioning. When placed under mental strain, as when operating a mobile phone, drivers decrease their speed and increase the standard deviation of lateral positioning on the road (Haigney et al., 2000; Cnossen et al., 2000; Fairclough et al., 1991; Vollrath and Totzke, 2000). This effect is found in the use of both hand-held and hands-free mobile phones as they both place a significant workload on the operator. Salvucci and Macuga (2001) studied driver performance in relation to in-vehicle interfaces. They examined how manual versus speech interfaces affected the variance of lateral positioning. There was a difference between the use of manual and speech-based dialling systems during driving. Manual dialling resulted in higher deviations of lateral positioning than speech-based dialling. In terms of longitudinal vehicle control, Vollrath and Totzke (2000) carried out a simulator study where the subjects solved an adapted version of the Baddeley Working Memory Span Test while driving. The interface representations were manual, visual or auditory. The results indicated that manual interfaces result in poorer performance than visual and auditory interfaces in terms of lowered mean speed and increased standard deviation in speed.

In sum, the aforementioned studies indicate that manual interfaces place more workload on the driver than auditory interfaces, and lead to deterioration in driver performance. Measures such as NASA-TLX, and mean and standard deviation of speed and lateral positioning, substantiate this claim.

The type of interaction that the interface is mediating is also a matter of concern. Some studies have examined the effects of having conversations or solving problems while driving (Haigney et al., 2000; Vollrath and Totzke, 2000; Lai et al., 2001), while other studies have investigated how drivers perform while interacting with formal information systems. Lee et al. (2001) performed a simulator study where the subjects interacted with a speech-based email system while following a lead vehicle that braked erratically. It was found that the subjective workload increased as the complexity of the system increased. Complexity was defined in terms of the number of menus and number of options in each menu. Performance, measured as increased reaction time, deteriorated as complexity increased.

Formal information systems present special challenges in terms of interface design. In addition to the issue of the representation of information itself, the structure in which it is embedded within have to be coped with. The interface must provide the user with an ability to functionally relate to structure, that is, efficient interface navigation. Interface representations of in-vehicle information systems are thus a matter of going further than just eliminating conflicts between cognitive and physical demands in primary and secondary tasks. Based on various studies, one can generally assume that speech-based interfaces have a certain advantage over tactile, hand-held interfaces (GUI), yet there is a possibility that this does not hold when applied to formal informational systems, due to the special features of navigation. This objection is particularly important because formal informational systems involve a strong component of navigation (Woods and Watts, 1997), which is not present in other types of interaction.

In sum, empirical studies indicate that the type of interface (speech, visual, manual) as well as the nature of the interaction (normal conversations, problem solving, dialling, using a formal informational system) have consequences for the outcome of driver performance. Given the wide distribution of in-vehicle information systems, and that their use implies that drivers relate to information that might or might not be relevant to driving itself, the effect of different forms of interface representations needs to be explicitly studied in order to provide a safe and efficient traffic environment.

This paper reports specifically on how a traditional tactile interface on a mobile phone (a graphical user interface (GUI) and Wireless Application Protocol (WAP)) and a fully speech-based interface perform when applied to a formal, high complexity, informational system while driving. Different interfaces operate on the same system of high complexity; this combination of interface and type of interaction has not been thoroughly investigated within transport research.

Based on the empirical findings and arguments presented above, the following hypothesis is stated: the comparison of hand-held and hands-free interface representations applied to an in-vehicle information system will result in dissimilar driver performance. The task of obtaining information from the system will influence the task of driving by inducing elevated levels of workload imposing behavioural adaptation. In terms of objective measures of driving, drivers handling a tactile/graphical interface, as in a hand-held mobile phone, will perform poorer than drivers handling a fully speech-based interface (input/output). Further, both conditions will have indications of elevated workload in relation to a baseline run, with no secondary phone task.

2 Methods
2.1 Participants

The study sample consisted of 20 participants (16 males, 4 females, in the age group 21–32 years, with an average age of 26.4 years at time of study in a within-subjects design. All participants held a Norwegian driver's licence for four years or more. All were familiar with the use of mobile phones, some also with WAP (4 participants). All participants reported normal or corrected-to-normal vision, and normal hearing. Written consent was obtained from the participants, and they were paid for their participation.

2.2 The simulator

The driving simulator used in the study is based on a 1998 model Renault Megane with manual shifting. All original controls give input to the simulator. Sound effects imitate sound from the engine, the transmission, as well as other relevant background noises.

The graphics projected on the screens are powered by a Silicon Graphics Skywriter workstation. Five channels of visual information provide the Field of View (FoV). Each screen is 2.4 m high and 3.1 m wide. The resolution of the visuals is 1024 x 768 pixels. The three front screens are rear projected and provide in sum 180 degrees horizontal FoV and 47 degree vertical FoV. The two screens behind the vehicle are front projected and supply in sum 90 degree vertical FoV and 47 degrees vertical FoV each.

The scenario (i.e. the route which the participants were requested to drive along) consisted of rural highways as well as city streets with intersections, secondary roads,

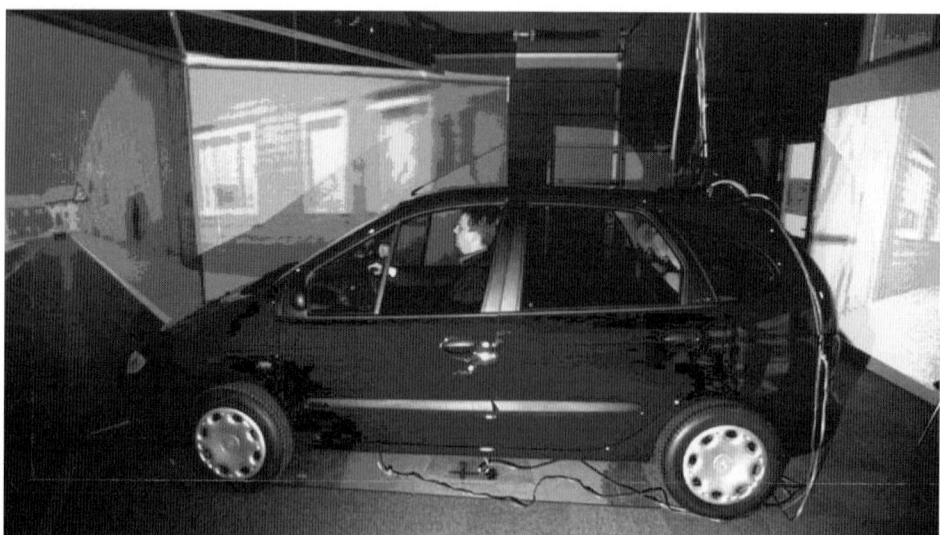

Figure 5.1 The simulator (photo by Jenssen, G.D.).

traffic lights, roundabouts, pavements, and buildings close to the road. The scenario starts in rural surroundings, on a road with one lane in each direction. After a few minutes driver enters a city environment. In the study, when entering the city, directions were given to the participants through the in-vehicle speakers. In order to provide experimental control, no external agents (such as pedestrians or other cars) were included in the scenario.

A mobile phone (off-the-shelf Nokia 6110 with hands-free kit) was mounted in the centre of the dashboard. An external microphone was attached to the phone, placed under the roof inside the car and aligned to the far left just above the level of the driver's face. The speaker was connected to the in-car stereo.

2.3 Procedure

The participants performed a total of four sessions, each lasting approximately 10 minutes: practice drive, two experimental trials, and a baseline trial. Upon arrival, the participants performed a practice drive in order to become accustomed to the simulator set-up. No task was given during the practice drive. The two experimental conditions were as follows: condition one included the use of a hand-held mobile phone, whereas condition two included the use of a speech-based interface. The third condition served as a baseline. The same route as for the experimental conditions was driven, but no tasks were assigned during the baseline trial. The experimental conditions were counterbalanced.

As a measure of subjective perceived workload, NASA-TLX (a subjective workload assessment tool) was administered after each driving session in the simulator (Hart & Staveland, 1988). The NASA-TLX was also administered before the practice drive during the introduction of the experiment, as an indication of the perceived workload of normal driving. The participants were asked to fill out the self-report form with reference to how they generally experienced driving in real life under normal, everyday conditions. Upon completion of the final session, participants filled out a debriefing questionnaire in addition to the NASA-TLX.

The task that the participants faced during the trials was to obtain various pieces of information from a formal information system. The information and its structural organisation contained in the database was equivalent, whether obtained by way of the WAP – interface or by way of the speech-based interface. Any difference in performance (i.e. driving) was hence due to the interface as such.

The experimental conditions were introduced to participants by informing them about the basic functioning of the speech-based and WAP interfaces, respectively. Dialling a number to the database-system activated the speech-based service. A spoken welcome message and the main menu indicated the activation of the system. The participants then navigated through the information structure by way of spoken commands (similar to those verbally presented in the speech-based system). Dialling a number that initiated an index WML (Wireless Markup Language) page containing a welcome message and the main menu activated the WAP service. The participants then navigated by activating the links in the various menus of the information structure. The WAP system made use of a traditional mobile phone interface (using mobile phone buttons for highlighting and choosing menu options from menus on the display). The participants were allowed to practice until they acquired a subjective minimum level of coping with the respective systems before engaging in experimental trials.

The respondents were given three tasks when handling the informational system. The same tasks were assigned for both experimental conditions. The first task was to report on the weather forecast for northern parts of Norway. The second task was to do the same for southern parts of Norway. The third task was to obtain the winning numbers for the national lottery and to have them sent to the mobile phone as an SMS (Short Message Service) message. In order to avoid having participants spend cognitive resources on remembering the tasks during the trial, the participants were provided with a piece of paper on which the tasks and their order were described. The participants were instructed to start performing the tasks at a particular reference point in the scenario, and were requested to drive as they would in a real-life setting. The participants were further instructed to terminate the session when all three tasks were completed.

2.4 Programming and data handling

The information structure in the speech-based service was implemented as wml-code (WAP). The hierarchical configuration of information was replicated in terms of the spoken menus being translated to text menus. The information content was also transformed to text. The links in the WAP replication were equivalent to the command words existing in the speech-based service. The WAP site was placed on an external server and functioned as a standard WAP site.

In this study driving performance was defined as average and variance of speed and lateral positioning on the road (e.g. the distance from the road centre). Both speed and lateral positioning were electronically recorded at 10Hz.

The log files from the simulator containing samples of speed and lateral position were transformed from .txt files to a file format available for SPSS for Windows. During the subsequent data analysis, T-tests performed on speed, lateral positioning and NASA-TLX were one-tailed in accordance with the specific hypotheses stated for this study.

3 Results

There was a difference between the hand-held condition and the speech condition on the effort and time pressure dimensions of the NASA-TLX.[1] Scores from the hand-held condition were higher than the speech condition (t = 1.94, df = 19, p < .05 for time pressure and t = 2.11, df = 19, p < .05 for effort) (Table 5.2) (Figs.5.3 and 5.4). The overall score did not reveal any significant difference between the two, but showed a moderate trend (t = 1.47, df = 19, p = .078). NASA-TLX subscales revealed significant differences between speech and baseline, and between hand-held and baseline (Figs. 5.2-5.5). The hand-held condition had a higher score in terms of being perceived as more demanding and having a higher workload) than the speech condition (for full statistics see Table 5.2).

Table 5.1 NASA-TLX average and SD of total score and sub-scales.

	NASA Total	Mental	Physical	Time Pressure	Effort	Frustration	Performance
Normal	32.12 / 12.13	41.10 / 19.10	19.90 / 13.85	33.70 / 15.90	28.60 / 19.98	36.20 / 18.44	33.20 / 14.07
Test	44.86 / 9.93	57.95 / 17.15	42.25 / 18.68	13.85 / 14.36	51.15 / 16.67	42.75 / 24.36	61.20 / 17.70
Hand	56.73 / 19.37	69.25 / 21.78	47.15 / 27.05	42.70 / 27.18	66.25 / 21.49	59.15 / 20.96	55.90 / 23.36
Speech	48.80 / 15.79	58.25 / 21.25	44.15 / 23.47	29.85 / 24.50	52.85 / 16.39	55.00 / 20.77	52.70 / 21.17
Baseline	26.68 / 17.16	31.35 / 21.48	25.50 / 23.32	14.95 / 14.53	27.10 / 21.28	25.75 / 21.17	35.45 / 18.45

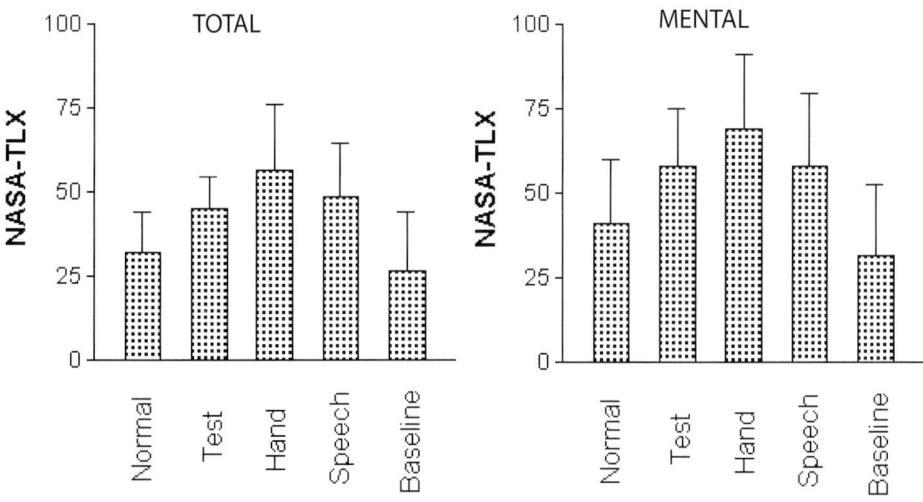

Figure 5.2 NASA – Total and NASA Mental subscales.

---

[1] Following Byers (1989), the NASA-TLX raw score rather than weighted score was used.

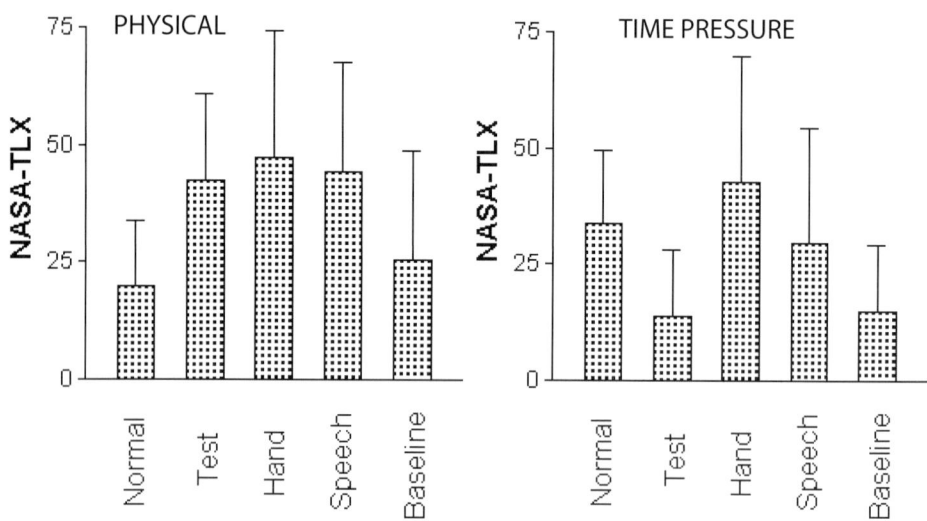

Figure 5.3 NASA Physical and Time pressure subscales.

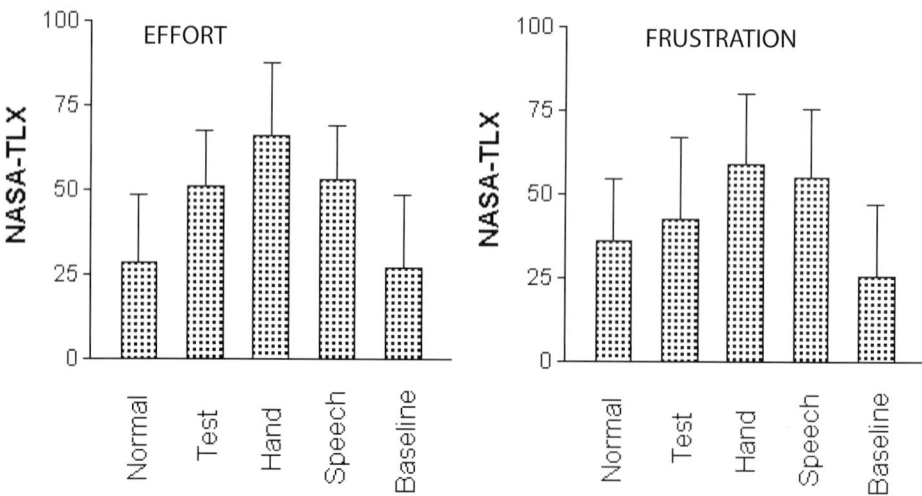

Figure 5.4 NASA Effort and Frustration subscales.

Table 5.2 T-values of NASA-TLX total and subscales: Mental load, Physical demands, Time pressure, Effort, Frustration, and Performance.

	NASA Total	Mental load	Physical demands	Time pressure	Effort	Frustration	Performance
Normal vs. test	−4.26***	−3.35**	−4.90***	4.15***	−4.46***	−.87	−7.99***
Normal vs. hand	−5.53***	−5.45***	−3.91***	−1.53	−7.23***	−3.47**	−5.14***
Normal vs. speech	−3.71***	−2.44**	−4.82***	0.69	−4.05***	−2.73**	−3.44**
Normal vs. baseline	1.02	1.37	−0.87	3.97***	0.21	1.50	−0.48
Test vs. Hand	−3.14**	−2.52	−1.21	−4.71***	−2.70**	−2.47**	0.99
Test vs. Speech	−0.90	−0.05	−0.35	−3.59**	−0.37	−1.57	1.57
Test vs. baseline	4.02***	4.36***	3.05**	−0.26	4.05***	2.52**	4.58***
Hand vs. speech	1.47	1.60	0.41	1.94*	2.11*	0.66	0.44
Hand vs. baseline	5.92***	5.98***	3.60**	4.75***	5.75***	5.31***	4.43***
Speech vs. baseline	5.24***	4.57***	3.26**	2.92**	5.21***	5.28***	2.96**

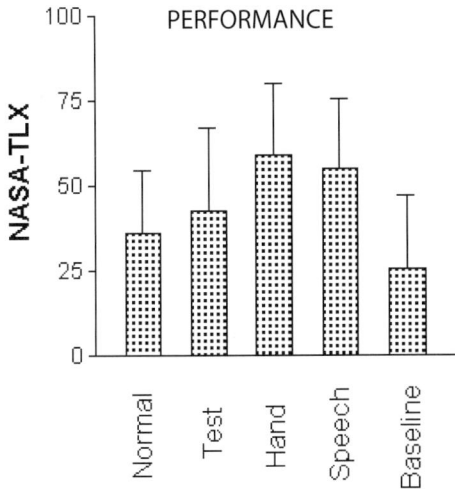

Figure 5.5 NASA Performance subscale.

The hand-held interface had a significantly higher overall workload than the test condition (t = −3.14, df = 19, p < .01). In particular, time pressure, effort and frustration contribute to this overall result. There was no difference, however, between the speech-based interface and the test condition (t = −0.90, df = 19, p > .05).

There was a difference in overall score between the perceived workload of normal driving and the use of the hand-held and speech-based interface while driving. Both experimental conditions scored higher on NASA-TLX (t = −5.53, df = 19, p < .001 for hand, t = −3.71, df = 19, p < .001 for speech).

There was a difference between the initial test condition and baseline, where the test condition elicited a higher score than the baseline condition.

There were no differences between scores on normal driving and the baseline simulator condition. This finding indicates that the simulator environment is ecologically valid for the assessment of the influence of representations in normal everyday driving (Table 5.2) (Figure 5.2).

As for average speed, differences were found between hand-held and baseline (t = −2.92, df = 19, p < .05), and between speech and baseline (t = −2.98, df = 19, p < .05). In both experimental conditions, subjects maintained lower velocities than in the baseline condition (Table 5.4).

There was no difference in average speed between the two experimental conditions (t = .51, df = 19, p > .05). There were no differences in SD between the two conditions (e.g. the degree to which there is variance in speed around mean). No differences were found either for average lateral position or for variance of lateral position in any of the conditions (Tables 5.3 and 5.4).

That the standard deviation in speed of the baseline is higher than for hand-held and speech interfaces, runs counter to the assumption that a high standard deviation expresses cognitive load.

In order to simplify the experimental set-up and reduce the risk of effects of motion sickness and/or dropout, the baseline condition was not counterbalanced, but was obtained after the second experimental trial. This allowed fewer rotations of experimental conditions in order to counterbalance (two experimental groups instead of six). However, the risk is that there could be a learning effect present in the data regarding the fact that baseline was always driven as the last condition. To account for this, an independent samples t-test was performed, including a Cohen's d value as shown in table 5.5.

Table 5.3 Mean and SD of speed and lateral position.

	Mean	SD	Mean	Sd
Hand	33.58	16.49	−1.10	0.95
Speech	33.53	16.39	−.99	0.94
Baseline	37.72	17.47	−1.12	0.93

Table 5.4 T-values for pairwise t-tests.

	Mean	SD	Mean	Sd
Hand vs. Speech	0.51	0.16	−0.89	0.05
Hand-held vs. Baseline	−2.92*	−2.04	0.58	0.33
Speech vs. Baseline	−2.98*	−1.59	1.07	0.14

Table 5.5 Driving performance during speech-based condition (N = 20).

	Speech-based first (n = 10)		Speech based second (n = 10)		t	Cohen's d-value
	Mean	SD	Mean	SD		
Mean Speed	33.69	4.84	33.36	3.55	0.175	0.08
SD Speed	18.03	2.75	14.76	2.22	2.919**	1.38
Mean Lateral position	1.10	0.13	−0.89	,75	-0,876	-0,41
SD Lateral position	1,03	,25	,86	,19	1,566	0,74

Table 5.6 Driving performance during hand-held condition (N=20). Independent samples t-test.

	Hand-held first (n =10)		Hand-held second (n = 10)		t	Cohen's d-value
	Mean	SD	Mean	SD		
Mean Speed	31,29	4,75	35,87	3,83	-2,38*	-1,12
SD Speed	15,43	2,79	17,54	3,22	-1,57	-0,74
Mean Lateral position	-1,09	,16	-1,11	,21	0,32	0,15
SD Lateral position	,99	,18	,90	,18	1,04	0,49

For the speech-based condition, as compared to the baseline, there is no difference with respect to mean speed (Cohen's d-value 0.08). This indicates that there are no training effects present. There is, however, a difference with respect to SD of speed, although it is difficult to interpret this as a general training effect (see table 5.6).

As for mean speed, however, there appears to be training effects present (Cohen's d-value −1.12). This indicates that some of the negative effects of the hand-held user interface might be due to training.

4 Discussion

Both the hand-held and the speech-based interface did impose significantly on both speed and workload in relation to the baseline. Hence, the use of mobile phones with both these interfaces increases workload and decreases driver performance compared to normal driving.

The data suggest, however, a better performance of the speech-based interface over the hand-held interface. In direct comparison between the two, the hand-held interface reveals a significantly higher load than the speech-based interface on time pressure and effort. The mental workload category shows tendencies, though not statistically significant, in favour of the speech-based interface. Additionally, the hand-held interface gives a significantly higher load on total workload score in relation to the practice drive condition, whereas there is no such difference between the speech-based interface and the practice drive condition.

As for the behavioural dimensions, differences were found between hand-held and baseline and between speech and baseline for average speed (but not for lateral positioning). Both experimental conditions showed lower speeds. However, it should be noted that there was a training effect present for the hand-held condition, and that this might have influenced the significant difference between hand-held and baseline. There was no difference in average speed or standard deviation between the two conditions per se.

No differences were found between any of the conditions regarding lateral position. Based on video observation, the reason for this seems to be that the participants tended to put the task aside when difficulties such as turns, roundabouts and crossroads occurred in the scenario. Hence, for straight roads this measure does not adequately discriminate between the experimental conditions.

No differences in workload were found between the participant's perception of workload of normal driving and the experimental baseline. This result might indicate that controlled experiments in a realistic and high-power driving simulator provide a sufficient level of ecological validity in order to make generalisations to normal driving.

There are indications from this study that the speech-based interface outperforms the GUI-based interface, but only to some extent. The main difference between the hand-held device and speech-based interface tested in this experiment might be due to the fact that the speech-based interface requires no motor component. Still, the results indicate that both experimental conditions proved to be significantly more demanding than the baseline, but not in relation to each other. Due to the fact that users are engaged in the activity of driving, certain resources are preoccupied (vision, hand and foot movements, etc.). When adding a secondary task (employing a mobile phone), the motor component of hand-held devices clearly presents a conflict in relation to central features of driving (steering the vehicle). The advantage of a hands-free operation of speech-based interfaces is thus quite obvious. This can be described by Baddeley's (1986) model of working memory, where separate mental resources process auditory and visuospatial information. Hand-held devices and driving both tap the same resource, whereas speech taps a separate resource. Yet, when interacting with an informational system requiring the driver to cope with navigation through menus and remember content given verbally, the advantage of a hands-free interface disappears.

The analysis of this study compares the scores on NASA-TLX (total and subscales) across five conditions (normal, test, hand-held, speech, and baseline). The high number of T-tests suggests a need for controlling the overall error rate. This may be done by downward adjustment of the alpha-level by way of a Bonferroni transformation of significance levels. The transformation for the analysis performed in this study specifies a corrected Alpha level of $p = 0.001$ (alpha $= 0.000714$). Applying this value to the set of data weakens some of the results in Table 5.2, where only p-values < 0.001 should be considered as statistically significant. There are several consequences of the Bonferroni transformation for the interpretation of the data of this study. Firstly, the difference in NASA-TLX scores (in the effort and frustration dimensions) between the experimental conditions loses its statistical significance. Secondly, the significant difference between experimental conditions and baseline in the dimension 'Physical strain' is invalid. Further, scores on 'Time Pressure' and 'Performance' for Speech are no longer significant in relation to baseline. In terms of objective measures, the analysis stating the significant difference between average speed between experimental conditions and baseline dissapears.

There is, however, some debate over the legitimacy of the Bonferroni transformation. Doing so, one decreases the risk of committing Type-I errors, but one also increases the risk of Type II errors. Some authors thus regard such transformations of alpha levels as flawed (for a full discussion see Perneger, 1998). For the analysis of this paper, the authors have chosen to keep the original alpha levels, yet being aware of the consequences in case of possible Bonferroni-transformation. The T-tests yielding p-values above a corrected level ($p < .001$) should thus be read with some caution. Still, the pattern of the results holds. The hand-held and speech-based interface induces higher levels of load than baseline, and measures of average speed show lowered speeds when interacting with an in-vehicle information system. Further, the speech-based interface does not fully outperform the hand-held interface, which is somehow in contrast to expectations based on other studies (e.g. Vollrath and Totzke, 2000; Salvucci and Macuga, 2001). Lee et al. (2001) imply that the complexity of the formal in-vehicle system places special demands upon the driver which the interface more or less support. It is quite conceivable that the nature of interaction (intentional acts in a formal complex system) calls for further explorations of speech-based interfaces. An alternative is the combination of speech with other types of interface solutions (Hoff, Alsaker and Bjørkli, 2002).

4.1 General discussion

The question of design of interfaces of in-vehicle information systems as a means of maximising efficiency and safety is particularly tricky in that it regards not one, but several issues concerning human cognition and behaviour. Additionally, the complexity of the problem gives rise to issues that do not surface in traditional cognitive science experiments. In addition to well-known facts relating to the cognitive and motor capacities of humans (e.g. theory of attention, signal detection), the task of driving is a real-time event, which in order to have applied value needs to be viewed as an activity.

To view a process as an activity, one has to incorporate the full range of dynamic and temporal aspects of the task. Activity always needs to be analysed in situ, because it is the process as such that is the main target of analysis. What is often found in unconstrained, descriptive studies of naturalistic behaviour is that behaviour is adaptive.

The results of the present study clearly show that the participants tended to adapt their behaviour according to the situation at hand. For both interfaces there was a tendency among the participants to put the task on hold during periods of high stress. This load-management is very important, because it is likely to occur as a phenomenon independently of other variables that strongly affect driving, e.g. skill, competence, attention, etc. If load management is a psychologically real phenomenon one would expect differences between interfaces on objective measures of driving to be evened out, because each driver would compensate, for example for lack of control, by simply waiting for the right moment to carry out the task.

It is also likely that skills-, rules-, and knowledge-based behaviour is mediated in the process of driving. These compensatory behaviours can be viewed in the light of the tendency of people to divide activity into distinct phases. There are three iterations: an orientation phase, an executory phase and a control phase (Leplat,

1991). Behaviour thus unfolds through chronological iterations where each phase has a subgoal of temporary organisation. The SRK taxonomy may, in this sense, not only refer to how the driver processes information, but also to how the driver organises his or her behaviour. This means that the levels of cognitive control are applied to the organisation of the activity of driving in terms of its phases/iterations. An example of activity phases from this study is observation of how the driver first performed navigational acts to find the activation point of the desired information, and then timed the activation and processing of it to low-demand parts of the run. This clearly expresses phases of orientation, execution and evaluation. An important aspect of a workload is thus how the driver chooses to tailor and organise the task. It is well known that users informally tailor the design of devices and work practices as a compensatory behaviour (Cook and Woods, 1996; Vicente et al., 1996). In line with the distributed cognition approach (Hutchins, 1995), it is tempting to suggest that one of the main cognitive contributions of the driver is to tailor the task and the artefact in a way that maximises efficiency and safety in the face of a continuously changing environment. How the driver chooses to tailor the task is an interface design challenge, in the sense that the interface must support the driver's need for breaking down activity into phases. By doing so, the interface supports the organisational efforts in which task demands are coped with in accordance to demands stemming from the task of driving. These organisational efforts are expressed in the three cognitive control modes in the SRK taxonomy, where the interface must have affinity for all three modes. Literature reviews have revealed that a) lower, skills-based levels of cognitive control tend to be executed more quickly, more effectively, and with less effort than higher levels, and b) that people have a definite preference for carrying out tasks by relying on lower levels of cognitive control, even when an interface is not designed to support this type of behaviour (Vicente and Rasmussen, 1992). For instance, in a stressful situation, in which the driver experiences little or no control, he/she is likely to adhere to rule-based behaviour, with little or no ability to pace the behaviour in accordance with the preferred tempo or in a safe manner. In situations in which the driver experiences a high degree of control, he/she is more likely to make higher-order or strategic decisions and set the pace of behaviour at his or her desired level (Hollnagel, 2000).

The small GUI-based in-vehicle information systems do not seem to support lower levels of cognitive control. On the contrary, they facilitate reasoning based on internal mental representations of the syntactic navigational structure of the interface. In particular, two factors contribute to this fact. First, the small size of such devices in general forces the interface to have a deep navigational structure. The number of functions and informational elements by far exceed the number of controls. Second, also to some degree a result of the size, the input devices tend to contain generic controls. Hence, much cognitive effort is put into operating the interface, rather than on the task itself. This means that before the operator can even start to think about the actual information required, he or she must work out how to relate to the structure in which the information is embedded.

When it comes to the speech-based interface, one may intuitively imagine that it facilitates an interaction style, which rests less on the pure syntax of the interface and more on the easiness of everyday conversation. This intuition is, however, somewhat mistaken. There are several obvious facts which indicate that speech-based interfaces and natural language invites to completely different types of action. In natural language, conversation occurs between two competent partners. The relevant competence in question is the seamless integration in a language of semantics and syntax in which the navigation is replaced by meaning. This implies that by expressing a meaningful phrase during a conversation, the conversational partner adapts and responds. In conversations, we navigate by intention and meaning. Speech alone does not imply conversation. To claim otherwise would be to mistake the act of speaking for the conversation itself.

In speech-based formal systems, the user adapts to the fixed informational structure by mentally representing an abstract fixed structure in which the relevant information is enclosed. Thus, intention must be transformed to instrumental commands to allow the user to activate the required information. To functionally relate to the system, the operator is forced to mentally map the relation between the syntax of the informational structure and the semantics of the device, as such. The informational device has no reference to the structure of information; the structure appears as a phenomenon on its own, separated from the generic controls of the device. In a formal informational system, there is a gap between the represented, and the structure it is represented within. When the represented is stripped from contextual constraints, the issue of navigation occurs (Woods and Watts, 1997). How to interact with the system is alienated from the reason why to interact. In other words, there is a clear-cut separation between structure and its content. To offer a speech-based interface does not do away with this fact.

The characteristics of formal informational systems accentuate the need for interfaces which more strongly supports the user in having to relate to both structure and content at the same time. The primary goal is to design an interface that supports several cognitive modalities, in which strain is distributed widely among available resources. This would further allow the user to work within all of the SRK behaviours, ranging from skills- to rules- and knowledge-based behaviour. This applies to both handling information as well as how to relate to the informational structure.

The SRK taxonomy is a factor that comes in addition to the particular resource that is in use. The choice of speech or hand-held interfaces is a question of what resources to tap. How this load is dealt with, is a matter of design. Visual, motor and auditory stimuli can all be understood as signals, signs or symbols, thus supporting a cognitive control mode, accordingly. In sum, it can be stated that both interfaces evaluated in this study suffer from the fundamental problem of having a wedge drawn between the representation of information and the representation of the informational structure. The larger this gulf, the more the interface is knowledge-based, and the less it is likely to improve on the efficiency and safety of the device at hand. This view is supported by the fact that the difference between the two interfaces examined in this study is somewhat marginal.

5 Conclusions

This paper has explored how two types of interfaces perform in relation to interacting with a complex formal informational system while driving. The main finding is that the hand-held interface does not trigger a significantly higher subjective workload than the speech-based interface. Both interfaces scored higher on workload measures in comparison to the baseline condition with no tasks. This result pattern is reflected by the objective measures such as average speed, where the levels of speed were higher in experimental conditions than baseline. The remaining objective measures (standard deviation of speed, mean and standard deviation of lateral positioning) showed no difference within or between experimental conditions and baseline.

The findings of the study show that when drivers interact with complex formal information systems, speech-based interfaces do not outperform the standard hand-held interface. There is a need to further explore interface solutions in order to provide viable ways of safely integrating formal information systems in vehicles and transport.

That many of the cognitive challenges of operating in-vehicle information systems regard compensatory activities (such as e.g. load management), rather than information processing in the traditional sense of the term, indicates that effort is needed to make such systems more flexible. It is the challenge of interface designers to develop systems that are based on the knowledge of the distributed nature of human cognition. This design process needs, as shown in the previous discussion, to be based on careful observation and analysis of the task, or rather, the activity of driving. Knowledge regarding the activity is the key to bridging the gulf between the representation of information and the representation of the informational structure. When this is achieved, the interaction will draw less heavily on knowledge-based behaviour and more on skills-based behaviour.

An additional means of supporting skills-based behaviour is to increase the perceptual ecology of the interface. In particular, it is important to increase the perceptual richness of interfaces, to develop analogue controls and feedback, to increase the semantic mapping between controls and their functions, to ensure an isomorphic mapping between input and output, and to ensure an ecological scaling of the interface devices (Hoff, Øritsland and Bjørkli, 2002).

For an example of an alternative representational interface for in-vehicle information systems in terms of some of the issues regarded in this discussion, see Hoff, Alsaker and Bjørkli (2002).

References

Alm, H. & Nilsson, L. (1995) The effects of a mobile telephone task on driver behaviour in car following situation. *Accident Analysis and Prevention*, 27(5), 707-715

Baddeley, A.D. (1986). *Working Memory*. Oxford: Oxford University Press.

Bateson, G. (1972). Steps to an ecology of mind. Chicago: The University of Chicago Press.

BSi (1996). Guide to in-vehicle information systems. *(Draft for development DD235: 1996). British Standard Institute.*

Cnossen, F., Meijman, T.F., & Rothengatter, T. (2000). Adaptive strategy changes as a function of task demands: a study in car drivers. *Submitted.*

Cook, R.J. & Woods, D.D. (1996). Adapting to new technology in the operating room. *Human Factors,* 38 (4) 593-613.

Fairclough, S.H., Ashby, M.C., Ross, T. & Parkes, A.M. (1991). Effects of handsfree telephone use on driving behaviour. In: *Proceedings of the ISATAS international symposium on automotive technology and automation* (p.403-409).

Fleming, J., Green, P. & Katz, S. (1998) Driver Performance and Memory for Traffic Messages: Effects of the Number of Messages, Audio Quality, and Relevance. *Technical Report UMTRI-98-22,* Ann Arbour, MI: The University of Michigan Transportation Research Instiute.

Goodman, M.J., Tijerina, L., Bents, F.D & Wierwille, W.W. (1999). Using cellular phones in vehicles. *Transportation Human Factors,* 1, 3-42.

Hart, S.G. & Staveland, L.E. (1988). Development of NASA-TLX (Task Load Index): Results of Empirical and Theoretical Research. In: P. A. Hancock and Meshkati (Eds.): *Human Mental Workload.* Elsevier Science Publishers.

Haigney, D.E., Taylor, R.G. & Westerman, S.J (2000). Concurrent mobile (cellular) phone and driving performance: task demand characteristics and compensatory processes. *Transportation Research Part F: Traffic Psychology and Behaviour* (p.113-121).

Hancock, P.A & Chignell, M.H. (1995) On human factors. *Global Perspectives on the ecologies of human-machine systems.* Vol 1. Lawrence Erlbaum Associates, publ. Hillsdale, New Jersey.

Hoff, T. (2004). Comments on the Ecology of Representations in Computerised Systems. *Theoretical Issues in Ergonomics Science.* 5(5): 453-472

Hoff, T., Alsaker, M. & Bjørkli, C.A. (2002). Effects of Tangible User Interfaces (TUI) in In-Vehicle Information Systems. In: Alm, H. (Ed.): *Proceedings of the Nordic Ergonomics Society's 34th Annual Congress on Humans in a Complex Environment.*

Hoff, T., Øritsland, T.A. & Bjørkli, C.A. (2002). Ecological Interaction Properties. *Proceedings of Nord-Design 2002.*

Hollnagel, E. (2000). Modelling the orderliness of human action. In R. Amalberti and R. Sarter (Eds.): *Cognitive engineering in the aviation domain.* NJ: LEA.

Hutchins, E. (1995). *Cognition in the Wild.* London: MIT press. Kaptein, N. A., Theeuwes, J., & van der Horst, R. (1996) Driving simulator validity: Some considerations. *Transportation Research Record 1550,* 30-36 National Academy Press, Washington, D.C.

Keskinen, E. (1996). Why do young drivers have more accidents? *In: Junge Fahrer and Fahrerinnen / Young drivers.* In: Berichte der Bundesanstalt fur Straenwesen, Mensch und Sicherheit, Heft M 52. Bergisch Gladbach, Germany.

Lai, J.,Cheng, K., Tsimhoni, O., Green, P. (2001) On the Road and on the Web: Comprehension of Synthetic and Human Speech While Driving. In *ACM SIG- CHI 2001 Conference Proceedings* (pp. 206 – 213).

Lee, J.D., Caven, B., Haanke, S., Brown, T. (2001). Speech-based Interaction with In- vehicle Computers: The effects of Speech-based E-mail on Drivers' Attention to the Roadway. *Human Factors.* 43(4) 631-640.

Leplat, J. (1991). Activités collectives et nouvelles technologies. *Revue Internationale dePsychologie Sociale,* 4, 3-4, 337-356.

McKnight, A.J. & McKnight, R.J. (1993) The effect of cellular phone use upon driver attention. *Accident Analysis and Prevention,* 25, 259-265.

Michon, J.A. (1989). Explanatory pitfalls and rule-based driver models. *Accident Analysis and Prevention* 21, 341-352.

Perneger, T.V. (1998) What's wrong with Bonferroni adjustments. *British Medical Journal.* 316, 1236-1238:

Rasmussen, J. (1983). Skills, rules and knowledge; signals, signs, and symbols, and other distinctions in human performance. IEEE *Transactions on Systems, Man, and Cybernetics*, SMC-13, 257-266

Redelmeier, D.A. & Tibshirani, R.J. (1997). Association between cellular and telephone calls and motor vehicle collisions. *The New England Journal of Medicine*, 336, 453-458.

Salvucci, D.D. & Macuga, K.L. (2001). Predicting the effects of cell-phone dialing on driver Performance. In: E. M. Altmann and A. Cleeremans (Eds.) *Proceedings of the 2001 Fourth International Conference on Cognitive Modeling.* (pp. 25-30)

Stevens, A. & Minton, R. (2001). In-vehicle distraction and fatal accidents in England and Wales. *Accident Analysis and Prevention*, 33, 539-545.

Vicente, K.J., Burns, C.M., Mumaw, R.J. & Roth, E.M. (1996). How do operators monitor a nuclear power plant? A field study. In: *Proceedings of the 1996 American Nuclear Society International Topical Meeting on Nuclear Plant Instrumentation, Control and Human-Machine Interface Technologies* (p. 1127-1134). La Granga Park, IL: ANS.

Vicente, K.J. & Rasmussen, J. (1992). Ecological interface design: Theoretical foundations. *IEEE Transactions on Systems, Man, and Cybernetics*, SMC-22, 589-606.

Violanti, J.M. & Marshall, J.R. (1996). Cellular phones and traffic accidents: An epidemiological approach. *Accident Analysis and Prevention*, 28, 265-270.

Vollrath, M., & Totzke, I. (2000). In-vehicle communication and driving: an attempt to overcome their interference, [Internet]. Driver Distraction Internet Forum sponsored by the United States Department of Transportation, National Highway Traffic Safety Administration (NHTSA). Available: http://www-nrd.nhtsa.dot.gov/departments/nrd-13/driver-distraction/Papers20033.htm #A33

Wickens, C.D. (1991). *Processing resources and attention*. In D. L. Damos: Multiple-task performance. London: Taylor and Francis.

Woods, D.D. & Watts, J.C. (1997). How Not to Have to Navigate Through Too Many Displays. In Helander, M.G., Landauer, T.K. and Prabhu, P. (Eds.) *Handbook of Human-Computer Interaction*, 2nd edition. Amsterdam, The Netherlands: Elsevier Science.

6

Theory and Practice of Ecological Interfaces: A Case Study of a Haptic In-Vehicle Audio System Design

Thomas Hoff, Hans Vanhauwaert Bjelland and Cato A. Bjørkli

> *Knowledge regarding fundamental invariant facts about user interfaces that are stable across culture, gender, expertise, age, domain knowledge, etc. is equally important as knowledge regarding aspects of the work system that are highly dependent on context. By reporting a practical interaction design project (the development of a novel interface for an in-car audio system design), this paper aims to demonstrate that knowledge regarding objective properties of the human/interface system is a strong prescriptive design tool in creating novel, exiting, fun, effective, and safe user interfaces. It has to be coupled, however, to sound User-Centred Design (UCD) methodology, in order to ensure these aspects are particular to a given domain.*

1 Introduction

What is an optimal user interface, in terms of effectivity, safety and enjoyability? This central question lies near the heart of human-computer interaction (HCI) enterprises. The obvious answer is, of course, that it depends. Yet what does it depend *on*? Does it depend on the technology involved? Does it depend on the usability of products as measured by heuristics and usability testing? Does it depend on knowledge of the cognitive capabilities of the users? Does it, on the contrary, depend on knowledge of the *activity* the product is situated within, i.e. culturally dependent factors? Alternatively, could it be that the design sensitivity of interaction designers is what matters the most?

Heuristics and guidelines cannot represent the full answer. They are after-the-fact abstractions that cannot be used predictively (up-front) in order to create optimal interfaces. To state that it can create optimal interfaces would be similar to stating that a painter could use the features of a famous picture, such as its size, color, brightness, and so forth, as a means to create a new picture. This is, of course, impossible. The value of heuristics and guidelines might have some function in evaluating the very surface of interfaces (the "touch and feel of interfaces"), but it is difficult to apply these principles up-front.

Many state that the quality of a design depends on the design sensitivity of the designer (i.e. his or her *expertise*). It is probably true that the accumulated experience of a designer is reflected in the quality of a design. However, it is conceivable that the 'magic' step from analysis (the knowledge of user, task and context) to design (the final solution) depends to a large extent on the quality of the analysis. The better the analysis, the less designer sensitivity will be needed in order to arrive at an optimal design solution. What matters is not necessarily the amount of analysis, but the soundness of approach. This is, of course, tightly linked to methodology. User-Centred Design (UCD) is a well documented method for moving from analysis to final design ((ISO 13407).

In addition to using UCD methods, the design process needs to be rooted in a theory about human functioning (representing the H of HCI), or more precisely, a theory of humans in relation to their external surroundings. There is a very good reason why the process should be rooted in theory of human functioning, and not in technology per se. Namely, we can make major changes in technology, but unfortunately we can do very little to change humans. Fifty years of research within the cognitive sciences have failed to find principal changes in the cognitive apparatus of humans, except a somewhat better ability to rotate images among men than among women (Sternberg, 2008). It is not entirely clear, however, what it really should mean to root the design process in a theory about human functioning.

How one chooses to describe human functioning clearly depends on one's methodological or even ontological conviction. It is interesting, but not very surprising, to find that methodology for describing the 'H' in HCI literature, co-varies with disciplines which have the sole purpose of describing 'H', most notably psychology, philosophy and anthropology. All of these disciplines have had an internal clash between positivistic and postmodern methodology (see e.g. Gardner, 1985). The same clash, although with a considerable time lag, can be observed in the HCI literature. There are very many examples of this, but Card et al. (1983), and Nardi (1996) (among others) stand out as strikingly different with respect to theory, methods, results, and generalizations.

In the transition from normative approaches to HCI (how workers *should* behave, based on rigorous modeling of operator and task) to descriptive approaches (how workers *actually* behave), the former stance has been heavily criticized for the notion that objective facts exist regarding user interfaces. The important point made has been that the most important processes in the work environment are not what happens inside the head of the individual worker, but in the context-based activity of the work system as such (because operator behavior is context based it cannot be predicted). Although the descriptive approach is probably right in asserting that there are no objective facts about user interfaces in the sense that not all aspects of the system can be modeled, but this is not to say that there are no objective facts about user interfaces *at all*. Every practitioner needs to be aware of both sides of the coin: a) there are fundamental invariant facts about user interfaces that are stable across culture, gender, expertise, etc., and b) there are aspects of the work system that are highly dependent on context, culture, gender, expertise and so forth.

Today, much is known about how to study b – in particular, the activity theory tradition has developed very useful tools (see e.g. Nardi, 1996). We have elsewhere attempted to develop an account of b which is not based on classical cognitive psychology but on ecological psychology (Gibson, 1979) and second generation cognitive science (Lakoff and Johnson, 1999).

This radically different approach has a starting point that diverges sharply from that of cognitive science and analytical philosophy, in that it views the mind as embodied, rather than disembodied. Very briefly explained, this means that there is no sharp distinction between what is on the inside (the subjective) and what is on the outside (the objective), and that the categories arise not from inherent properties, but as an emergent property of the perception/action system in relation to the physical world.

Having laid the theoretical foundations of this alternative account of a elsewhere, we wish here to demonstrate how, in a practical, real-life setting, this approach can be applied to interaction design projects. In the search for a suitable case where the principles could be demonstrated, we looked for products with many users, which do not require particular skills up-front, and where the implications of poor design are high. The reasons for these criteria were to focus on products which seldom receive much attention from the cognitive engineering tradition and to choose a product where the relation between the theoretical approach and the practical design solution is very salient. For a more traditional cognitive engineering case, displaying the same principles, see the work of Hoff and Hauser (2008), which studies graphical representations in energy management systems.

One product, which is used by many, that causes an enormous amount of accidents every year, and which has a fairly poor interface, is the car stereo.

1.1 In-car audio systems

The first in-car audio systems were radios installed in the early 1920s. The skeptics soon claimed that the presence of a radio distracted the drivers and resulted accidents. Tuning them was said to take the driver's attention away from the road and the music could lull him or her to sleep. The radio manufacturers argued that the radios actually represented a safety feature by keeping drivers awake. In spite of the criticism and proposed bans on radios the car industry steered clear of any restrictions on car radios. Since then car and radio manufacturers have sought to remove the physical distraction of tuning the radio by introducing push-button presets and automatic tuning.

The radio was later supplemented with other sources of media, such as the cassette and the compact disk. Today, even other sources such as minidisks, digital radio and memory cards are added. In addition to the original on/off, tuning and volume regulations, modern systems enable the user to adjust the quality and physical balance of the sound. They also present a large number of extra functions for both the radio and other media. All this makes tuning the radio not the only for drivers anymore. The audio system has become a complex system intended to be used parallel with the even more complex activity of driving a car.

1.2 Epidemiology

Compared to other systems either already in use or expected to be implemented in cars shortly, the audio system can be said to have a modest complexity and to be relatively easy to use. Because of the uncertainties and underreporting, it is difficult to precisely indicate the number of crashes caused by people interacting with the audio system. Still, Dr. Michael J. Goodman, human factor research specialist for the National Highway Traffic Safety Administration, US (NHTSA), says his group conservatively estimates that c.150,000 crashes in the US each year are related to ordinary radio use, primarily because it demands the driver's attention (Kobe, 2000). With a total of over 6 million crashes in the US (NHTSA, 2001), this corresponds to more than 10% of the total crashes caused by distraction that were studied. This is also similar to the number Stutts (2001) estimates.

Driver distraction may be described as "any activity that takes a driver's attention away from the task of driving". An investigation of crash data shows clearly that any distraction has the potential to cause or contribute to a crash. Interaction with an in-car audio system will necessarily increase the load on one or more of the attentional resources and thereby reduce the available resources for the primary task performance. The NHTSA estimates that approximately 25% of police-reported crashes involve some form of driver inattention – the driver is distracted, asleep or fatigued, or otherwise "lost in thought". Since it is implausible that drivers linked to crashes are likely to always tell the truth about their inattention we can assume that such accidents are underreported. Some argue that these numbers are as high as 35–50%.

According to Ranney et al. (2000), the potential for distraction associated with any secondary task is also determined by the "willingness to engage" in that task. This factor refers to the conscious or unconscious decision processes involved in electing and carrying out secondary tasks while driving. The willingness to engage is a function of a multiplicity of factors, including driver (e.g. experience), vehicle (e.g. display design), situation (e.g. urgency), and task characteristics (e.g. ease of use). An important safety consideration is also the driver's "willingness to disengage" from the secondary task. It is the coincidence of driver inattention and the occurrence of unanticipated events that triggers distraction-related crashes. To be absorbed in a secondary task can be critical when a sudden change in the traffic needs more attention to be paid to the primary task of driving.

Drivers are distracted by things outside the car such as the scenery, and also mental elements such as something on their mind. However, research shows that over 50%

of the distractions that result in crashes are elements inside the car, such as objects, interacting with another person or animal, or interacting with instrumentation such as the audio system. As already noted, operating the audio system represents more than 10% of the distractions that cause crashes.

2 Analysis

In this section, we aim to present the two models applied in the design case study. The first section describes the human functioning theory, which is based on Hoff et al. (2002b). The second section describes the results of using traditional UCD methodology (an analysis of user, task and context).

2.1 Objective truth

We have stated that in order to arrive at an optimal user interface, the designer needs both to have a clear and explicit account of what objectively constitutes such an interface, as well as to use ethnographic methods in order to establish the context of the activity of carrying out a task (the latter describe factors that are context-specific, historic and dependent upon characteristics of the user and the task). In order for the designer to arrive at the former, e.g. a clear and explicit account of what objectively constitutes an optimal user interface, we present a theory based on ecological psychology.

Ecological psychology puts less emphasis on the internal mental mechanisms for the interpretation of incoming stimuli, and more on the interaction between the subject and its external surroundings. The reason for this is the assumption that man has evolved in a reciprocal relation to the environment. Because of this reciprocal evolution, there are seldom ambiguities as to the knowledge of things in the external, natural world (exceptions are some natural phenomena that give rise to illusions). Thus one can make empirical assumptions regarding the human perception/action system without referring to internal mental models, or even to neuronal activity.

When solving a task, explicit thinking arises only as a rupture in the perception/action system – it is the manifestation of a system breakdown. It can be brought about by carefully manipulating the task in order to make it difficult (such as e.g. the Tower of Hanoi problem). A paradoxical fact is that modern technology has an inherent tendency to create system breakdowns, even though designers aim to make products as easy to use as possible.

It is when the 'humanrange' (our unaided physical limitations of perception and action in time and space) is expanded with the use of tools (technology) that the problems start to occur (Hancock & Chignell, 1995). The aim of technology is usually to extend our 'prostheticrange' (e.g. the boundaries to aided action) and our 'universalrange' (the boundaries to aided perception). Designers usually perform well when their task is to create physical tools (such as prostheses, bicycles, helmets, etc.), but they often fail drastically when something is to be *re*presented. Every property that that undergoes a transformation from its original state must be represented.

Even though it might be difficult to detect, there is, in our perspective, a pattern that describes the difference between what is presented and represented, and between what is represented in a good manner and what is represented in a poor manner, and that explains the difficulty designers face when designing interfaces that are representations.

The reason why designers face greater difficulties when *re*presenting rather than presenting, and that there seems to be degrees of difficulty depending on the kind of representation they make, is due to the degree of disentanglement between the representation and the represented. We have elsewhere shown (Hoff et al., 2002b) that different ways of representing the same hammering device gives rise to very different phenomenal experiences, that neither depend on the technology involved in the representation (e.g. digital, electrical or pneumatic transmission), nor on interaction design labels (such as GUI, SUI or TUI), or solely on internal mental states. What they do seem to depend on is the degree of *directness* of the interface. Directness refers here to the degree to which invariants of the perceptual flow field are recreated or conserved in the final representation.

Although digitalization is a powerful and flexible means of treating information, it represents a case where there is nothing that constrains the relation between the representation and the represented. Hence this represents the maximum degree of disentanglement between the representation and the represented. Compare this with interfaces where there is a mechanical link between the representation and the represented. In this case, many of the perceptual invariants that are so important for humans in order to function optimally are conserved – not by the designer, but simply because of the nature of the representation. Woods and Watts (1997) has used the transition from old hardwired control rooms to full GUI control rooms to illustrate the same point. In the former case, the design of the system is fully visible because it has a physical layout in the control room. In the latter case, however, the functionality of the system is completely hidden behind the currently active window. What has been known as 'the keyhole effect' refers to the fact that using full graphical representations is much like looking at the world through a keyhole, where only a small subset of the total perceptual field can be seen at once.

The more disentangled the representation is from the represented, the harder it becomes for the designer to recreate the missing invariants. The more invariants that need to be recreated, the more the designer needs to know about invariants. The problem with this is that the empirical study of invariants is fairly novel; unfortunately, there is no 'Big Book of Invariants' on the market. So, the good news is that there exists an objective means of describing interaction – the bad news is that this knowledge is still in the making. In what follows, we will attempt to describe what we *do* know.

2.1.1 Ecological information

At a very general level, one of the most important insights from the ecological approach is the importance of invariance of space and time. This simply means that humans tend to perceive unfolding events as happening *in* time and *in* space. In natural occurrences, there are seldom any leaps of information, neither spatially nor chronologically. In computerized media, however, the lack of space/time continuity is the norm, rather than the exception (for a detailed discussion, see Woods & Watts, 1997; Hoff, 2004). Another important point stated by Gibson is to consider the senses as perceptual *systems*, rather than as separate sources of incoming stimuli. In natural settings, there is in this sense always an abundance of perceptual richness, where overlapping occurs between individual senses. Studies within the ecological tradition have shown that humans are actually capable of distinguishing the relative length of a stick only by the sound of the stick falling to the ground (Carello et al., 1998), and able to assess the relative length of a stick by the dynamic kinesthetic (the inertia tensor) of the stick when blindfolded (Santana & Carello, 1999). In general, very little emphasis has been put on the possibility of multimodal feedback (there are, of course, many individual exceptions to this fact). One of the reasons might be the lack of a unifying theory that describes the mechanisms of such an enterprise. The ecological psychology tradition is a conceivable candidate to offer such an account.

2.1.2 Skills and knowledge

The division between skills and knowledge, and the process of transition from knowledge to skills, is important within ecological psychology as well as within traditions associated with phenomenology. Within Human Factors, the most important utilization of this distinction has been put forth by Rasmussen (e.g. Rasmussen 1983). It is generally agreed upon that skill-based levels of cognitive control tend to be executed more quickly, more effectively and with less effort than higher levels, and that people have a definite preference for carrying out tasks by relying on lower levels of cognitive control, even when an interface is not designed to support this type of behavior (Vicente & Rasmussen, 1992).

Whether an interface is acted upon in a skill-based or knowledge-based fashion does not depend fully on the interface as such; rather, it depends on whether the interface information is interpreted by the operator as signals, signs or symbols. Hence, the relationship between representations and cognitive control mode is not fixed. Representations may engage in several cognitive control modes, but do not exclude any of them. The very same display may be interpreted as a signal, as a sign, or as a symbol, depending on the intentions, expectations and expertise of the observer (Vicente, 1999). However, there *is* a fundamental ecology of representations in the sense that they favor and support some modes rather than others. A very central question, which has not been adequately covered in the literature, is which principles regulate the chances of an interface being interpreted as a 'signal' in Rasmussen's terminology. We have elsewhere (Hoff et al., 2002b) stated that there are properties of the phenomenal human/interface interaction that have an affinity of being interpreted as signals. The main distinctions in the EIP framework (Ecological Interaction Properties) will be considered next.

2.1.3 Ecological Interaction Properties

The degree of 'directness' of an interface is here determined by the smallest element to which a change in that element introduces a change in user experience. Because the properties represent degree of directness, they are continuums rather than discrete categories. The description of the EIP framework is identical to description in Hoff & Øvergård (2008, this volume).

Continuous-Discontinuous Property. This property relates to the representations of information presented in the interface, and to the degree that these representations are continuous or discontinuous.[1] This is related to whether the information is presented as continuous events that simulate the spatial and temporal aspects of real-life events or whether they are presented in a bit-by-bit fashion (Hoff et al., 2002b). It is assumed that normal supervision of the function of a given system probably will benefit from a dynamic continuous display of information (Hoff & Hauser, 2008). This property may be described as a continuum that at one end includes an interface that has continuous information representations, while at the other end it contains a static representation of the state of the system that changes like a photographic slide show.

Specific-Generic Property. This property relates to how many functions a given control has. The relation between the number of functions and the number of controls leads to two types of complexity. With many controls, each of which have one function, the *apparent complexity* (or perceived usability) of the interface will increase (or decrease, in terms of usability). When there are few controls, each with a greater number of possible functions, the *complexity of use* (or actual usability) will increase, since the increased complexity of the controls must be mirrored by a greater complexity of use by the operator (Norman, 2002).

Perceptual/Motor Scaling Property. This property relates to how the design of the interface is fitted to the human sensorimotor capabilities. It can be divided into two different aspects. These aspects are closely related to the use of anthropometric data and human factors guidelines, but are different in that they focus on the ecology of the total experience of interaction, which includes bodily movement and different sensory modalities (Hoff et al., 2002b). On the other hand, they differ from anthropometric data because they are not only anatomical but also functional. This is because the size of many environmental objects must be fitted not only to the size

---

[1] This continuum has been named Analogue-Digital by Hoff and colleagues (2002), but this name has led to difficulties since it has cultural connotations related to the form of the equipment rather than the nature of the representations in the system.

of the operator's body, but also to the task and the actions of the operator (Warren, 1995).

The first aspect is how the interface is scaled in regard to the limitations of the human sensory system: Does the interface contain (or present) light (lux and color contrast), sound (Hz, pitch, timbre, etc.), tactile stimulation (surface texture, etc.) and haptics (when manipulating controls, etc.) that are *within the perceptualrange* of the human operator?[2] An interface that induces direct interaction should be fitted to the biological limitations of the human sensory system, e.g. it should be fitted to the perceptualrange (Hancock & Chignell, 1995).[3]

The other aspect is how the interface and its controls are scaled to the human body and the movement pattern of the operator. This relates to the size, shape, weight, and surface structure of the interface and its controls. Research has been carried out on the ecology of human movement and action in an environment. Body-scaled biomechanical invariants have been found in stair climbing (Warren, 1984), when passing through apertures (Warren & Whang, 1987), in sitting height (Mark & Vogele, 1987), and in grip transitions (Cesari & Newell, 2000).

These invariants can be said to afford[4] specific types of actions from the observer (Gibson, 1979). By locating these body-scaled invariants that guide (and can be used to predict) human behavior, researchers can produce more ecologically correct surroundings (Warren, 1984). If it is correct to assume that such affordances exist in the natural world as perceived by humans, and that these guide our actions, it should not be too bold to assume that the use of these biomechanical invariants can also improve interface design.

Perceptual Richness Property. This property relates to how many senses are involved in the interaction with the interface (Hoff et al., 2002b). The human body, through co-evolution with the environment, is adapted to multimodal sensory input. The five perceptual systems (orienting, auditory, haptic, taste and smell, and visual) are interrelated, and they work as a system where they all contribute to the perception of an event or an object (Gibson, 1966). A multimodal and perceptually redundant event will involve several senses and as such also include several types of stimulus information that can be picked up by the observer (Gibson, 1966). If an event is perceptually 'poor' and only consists of one sensory modality, less information is available for the observer to pick up. This may lead to the need for additional stimulus information as the information from the event is too scarce and/or impoverished to be the basis for effective perception or recognition.

---

2   Olfactoric and gustatoric feedback are not mentioned here, as these sensory modalities are not equally important as the other senses when interacting with computer systems.
3   The perceptualrange is the part of the environment that the humans can perceive without the help of tools or technology, e.g. it is the range for unaided perception (Hancock & Chignell, 1995).
4   Affordances (a neologism first presented by Gibson, 1979) describe what can be done with a given object. They define the actions made possible for a given observer with a given object (Turvey et al., 1981). One can also say that an affordance specifies the action capabilities of an object, and can only be found on an ecological level (Gibson, 1979, see also Zaff, 1995).

Motoric-Sensoric Mapping Property. This level is related to the degree that the representations in a system are spatially related to the actions of the operator. The actions of the operator occur in a four-dimensional space[5] where objects generally conform to the size, speed and directions of these movements. Consider the example of moving a book, where the book held in the hand will conform in a one-to-one relation with the arm/hand's movements. It is not only in the physical world where there is conformity between the movement of the observer and changes in the world. This also seems to be the case with visual imagery, where researchers have found that the time taken to scan a visual image is correlated with the time taken to scan the actual object. It seems that visual images conserve the functional size of the imagined object (Kosslyn, Ball & Reiser, 1978). From this it might be argued that both the physical and mental movements of a person reflect, and are based upon, the relation between the world and the perceiving individual (for a similar argument, see also Lakoff & Johnson, 1999).

Input/Output Property. This property is related to the degree that the operator perceives a gap between the input given and the output received. This gap could either be a temporal or a qualitative difference between the input and the output. A temporal lag could be exemplified by an overloaded computer, which does not have the CPU capacity to handle all the information at once, something that leads to a delay in responses to the operator's commands (input). A qualitative difference could be that the output consists of sensory modalities that are not prospected by the operator, given the modalities of the input. Other qualitative differences could be a plane-mapping discrepancy, such as presented by the use of a mouse to move the cursor on a computer screen (Hoff et al., 2002b).

Physical-Inferential Property. This property is composed of three different aspects relating to the representations in the interface and the operator's actions and the consequences of these actions (or the representations of these consequences). These properties overlap to some degree and they are possibly not even exhaustive, but they will be described separately to better show the important aspects of this ecological interface continuum.

The *first* aspect is *prospection of consequence*. This is the semantic relation between the controls and their functions. This aspect relates to whether the consequences of the use of the control are mediated by the layout of the control. The first aspect can be described by the question: "In what way does the control's layout inform the user about the consequences of manipulating the control?" A direct interface in this respect is an interface where the outcome of manipulation is perceived without any cognitive inference. An indirect interface is one where the operator has to employ systematic problem solving before s/he can use the control effectively. The concept of prospective control (Hofsted, 1996; Lee, 1993) becomes important here, as the

---

5   The four dimensions are height, width, depth, and time. Four-dimensional space used to denote that the actions of a human are always spatially and temporally placed.

design of the interface can help the operator plan ahead when operating the system. As Hofsted (1996, p. 33, italics added) claims, "All actions are geared to the future and *controlling them requires knowledge of upcoming events*. We continuously need to know what is going to happen, both within the near and the far future, in order to plan our activities and to coordinate our movements".

The *second* aspect is *representation of consequences*. This relates to whether there is a *meaningful relation* between the *output* and the *representations of the output*. This aspect can be described by the question "Do the representations of the results of the machine's actions (given that the operation does not lead to physical results) relate in a meaningful way to the actual result?" A direct interface is one where the operator will know which operations the system has performed by viewing the representations of the results. These representations may, for example, be graphs, symbols or pictograms.

This problem is highly related to human-computer interaction, as the output in a computerized system is often of a digital, graphical nature. Physical, real-life devices often have direct output. A hammer, for example, has a direct physical output that includes several sensory modalities (haptic, visual and auditive senses), that stand in a natural relationship to the input (hitting with the hammer), whereas the consequences in a digital computerized system often have to be inferred as they are "hidden inside" the system, and the feedback representing the consequence often is in an abstract/symbolic relationship to the consequence. It is also possible that the system does not give any feedback to the operator.

The *third* aspect is *response mode guidance*. It relates to the degree that the control informs the operator about what type of response is needed to effectively manipulate the control. Should the control be pushed or turned, should it be pulled or pressed, which part of the body is it fitted to? This aspect is close to what Norman (2002) calls the "gap of execution", where the operator can perceive what actions are possible and what actions should be performed. This could be an example of what is normally called an affordance (Gibson, 1979). In a computerized system, many symbols or icons are arbitrary and do not afford the operator information as to how to manipulate them in a successful way. By simulating the natural affordances in a computer system, designers may improve the usability of these systems (Norman, 2002).

Certain qualifications need to be made concerning the ecological interaction properties. First, they overlap. The criterion for a property is that it should relate to a significant aspect of the ecology of a particular interface that is not fully covered by any other property or combinations of properties. The categorization of properties into the ecological interaction properties described is based as much on prototypes as on clear-cut lexical definitions. That is, there are prototype examples, that highlight salient differences between properties, but there are also instances where the difference between the properties becomes somewhat blurred.

Because there are similarities and cross-couplings between the properties, and because some properties might in certain cases be special instances of other properties, it makes little or no sense to treat the properties at a quantitative level. Rather, the aforementioned continuums should be viewed as a general qualitative tool for aiding the assessment of the degree of directness of a particular interface.

2.2 What is wrong with current interfaces according to EIP?

The shortcomings of the traditional in-car audio systems with respect to the EIP categories above are principal. We do not aim at an exhaustive analysis of the directness of this category of interfaces here, but some very influential shortcomings deserve consideration. First of all, their perceptual scaling is very poor. In relation to the primary task, which is driving, the input knobs are far too small. The same goes for the output display, which is also too small with respect to the fact that it is located far from the driver, and that it should be possible to assess the status of the device at a short glimpse. Many of the input knobs are 'digital', e.g. up-down and back-forth rather than continuous. The volume adjustment is a good example of this, as the continuous volume knob has been replaced with discontinuous up-down knobs. A further problem with the volume knob with respect to the EIP properties is that the input is further divorced from the output, in that one push on the volume knob gives little effect on the output. In general, it is a major problem that all of the visual output is presented on the device, which is very awkwardly placed with respect to the line of sight while driving. Further, with respect to the specific/generic property, many of the input controls are generic, rather than specific. Hence, mode errors are frequent. As for perceptual redundancy, most of the structural information is presented visually (although with respect to *content*, the feedback is auditive). Visual feedback is not very appropriate in this setting, due to the fact that driving is a very visual task in the first place. In particular, auditive cues and haptics are suitable candidates for augmenting perceptual feedback.

A novel attempt at designing safe and effective in-car audio systems should aim at eliminating the problems mentioned above. We have, however, no precise methodology for implementing the EIP properties in the final design. In this case study, the designers where introduced to EIP theory before the physical design work started, and the design solutions where iterated based on reference to EIP categories.

2.3 UCD Analysis

In this section, we describe a UCD analysis of the user, task and context implied in car driving and operation of in-car audio systems.

2.3.1 Background

Today the CD player and FM/AM radio are the main sources of media in the car. The cassette deck still exists in some audio systems, while new sources such as the minidisk and different types of memory cards are being adapted for use in cars. It is also not unthinkable that, as mobile phones are more integrated into the car by technology such as Bluetooth, they might also contain music. As other audio media like portable devices and computers develop, consumers will expect some of the same functionality in cars. The way we think about car media is likely to change in the years to come. New in-car audio systems will need to meet the possibilities and challenges this involves.

Some producers already include a reasonably priced hard drive, which, in the same way as a CD changer, allows the user to play a multitude of tracks without physically having to change the CD. The same possibility is given with the use of data CDs recorded with sound files compressed in formats such as MP3 or WMA. Furthermore, the car industry is working on equipping future cars with wireless network.

When only one album (CD or cassette deck) was used at a time, the term *track* was sufficient to describe what to play. With hard drives and CD changers, the concept of *album, media pack, folder,* or *play list* needs to be added. The one-dimensional "tracks on one album" has become a two dimensional "tracks on many albums".

DAB and SDARS are two new standards for radio transmission and receiving that is expected to manage market penetration within a couple of years. These systems offer better sound quality, more functionality and more available channels. Instead of frequencies, the user will relate to program numbers.

Even though automatic radio tuning makes it possible to use a radio without knowing the right frequency, the frequency is still a good help to find the right station if there are many stations to choose from. This might further be enhanced with the introduction of digital radio, where the user needs some sort of shortcut to quickly navigate to the right station among the approximately 200 stations there are to choose from. An alternative to using frequency is to let the system sort them, for example, alphabetically by name. Another way of handling the large number of

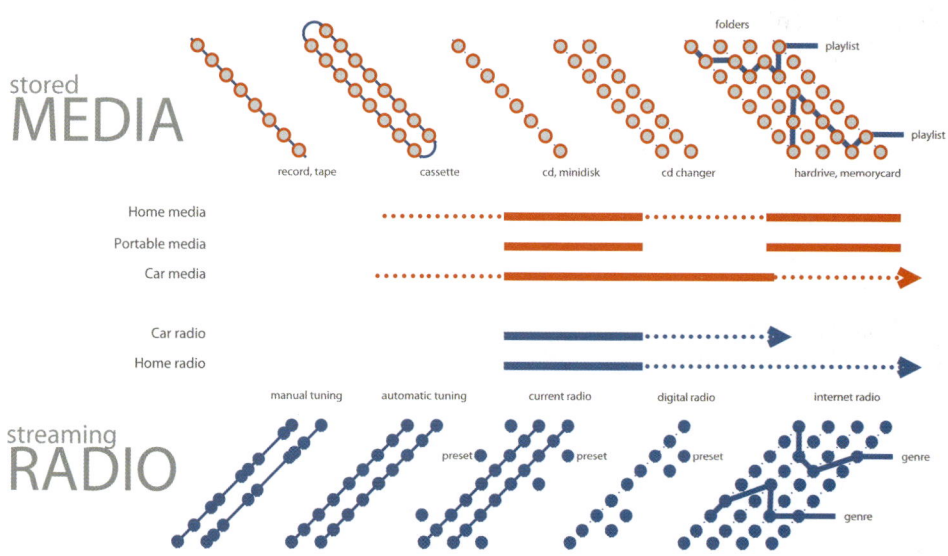

Figure 6. 1 Development of sound media. Future systems need to meet the development of technology, as well as the users' habits and expectations. The evolution of different areas is illustrated in the middle part.

channels is to group them into groups or *families* by some sort of system like genre or geography. The interfaces of today's audio systems are generally not scalable to handle a scenario of 200 stations in an adequate way.

2.3.2 Conventions

Most users are likely to have experiences from other audio systems before they use the new system. These systems are, on the whole, following interface conventions that have evolved through the history of systems such as hi-fi systems, portable players and in-car systems. Some conventions are standard for all audio systems while others are specific to a particular category. A new product or system can be elegant and brilliant, but if its interface violates strong conventions the users will have problems using or adapting to it. The question for the audio system is how strong the existing conventions are, and what the results will be if they are broken.

The most apparent convention is the placement of the in-car audio system, which is in the mid-console. More and more often on premium cars there are also control buttons on the steering wheel, controlling volume, and track and station change, for example.

It seems that the use of English text labels and standard signs is an accepted convention. A horizontal organization of controls is also standard. A rotary knob most often controls the volume, whereas a clockwise rotation increases the volume. The same knob often controls the 'off' function, either by pushing it or turning it counter clockwise until it clicks. The rotary knob convention is, however, not the only convention used. Some systems use two push buttons to increase or decrease the volume discretely. The bass, treble, balance (right / left balance of sound) and fader (front / back balance of sound) are controlled in a number of ways, such as individual rotary knobs, coupled multifunctional knobs and menu-based control.

2.3.2.1 Stored media

The navigational buttons for products such as the CD player are most often arranged horizontally in something like the following sequence: previous track, fast rewind, stop, play, pause, fast forward, and next track. These are symbolized by ⏮ ⏪ ■ ▶ ❙❙ ⏩ ⏭, respectively. Sometimes the ⏪ and ⏩ buttons are omitted, requiring the user to press and hold down ⏮ or ⏭ to fast rewind / forward. The ❙❙ button is occasionally included in the ▶ button, letting the button toggle between play and pause.

The loading of stored media sources as CD's, cassettes and minidisks is done "slot in" This means that they simply are led into a slot in the interface, in contrast to the CD tray on most computers and hi-fi systems, or the cassette or CD lid on portable devices.

2.3.2.2 Radio

To tune the radio some of the same physical buttons as for the CD are used. Tuning up or down is done by pushing the ◀◀ or ▶▶, or the I◀◀ or ▶▶I buttons. On earlier radios tuning was done manually with a rotary knob, typically on the opposite side of the display to the volume knob. This organization of two rotary knobs on each side of the display can still be seen in many audio systems today. Modern radios allow the user to store favorite radio stations using presets buttons (typically six) for later use. The way this is done varies greatly, but the most used methods involve hitting one of the preset buttons and holding it down for a period of time, or first hitting a memory button before hitting the preset button. The different methods of operation and the somewhat abstract concept of presets make the storing of presets a difficult task. Retrieving preset stations is easier and simply requires the user to press the preset button. Modern radios also have a function allowing the user to automatically search for the strongest stations and store these in the preset buttons. There is no symbol for this and the button has short forms such as "AST", "ASC" or "SCAN", which cause many users to have to look twice to find the right button on an unfamiliar system.

The RDS (Radio Data System) system, which works as an enhancement of the FM frequency band, allows radio stations send out digital information parallel to the sound. This makes it possible, for example, for the user to let the system change station whenever another station broadcasts traffic information. It also lets users search for stations broadcasting sports, jazz, news, or another particular program type. The support and buttons to control this functionality vary greatly. The most applied function is the 'Traffic information', most often operated by toggling the on / off state by pushing a button.

We see that there are significant variations between different in-car audio systems and also that users are accustomed to various interfaces. Still there are some established conventions and expectations relating to in-car stereo systems. Possibly the most notable ones are the placement, the presence of functions such as volume, next, previous, etc., and the structure of functions relating to source and track / station.

2.3.3 How do people use the audio system?

Numerous conversations and informal interviews with people with different backgrounds confirm what common sense has told us all along—that the large majority of users do not use advanced functions in the audio system. They are simply interested in easily being able to choose what to play and how (loud) to play it. Sophisticated functions such as the Program Type (PTY) of RDS radios, or even not so sophisticated functions such as the balance, fader, bass, and treble, are often reported to be left untouched after the initial experimental phase.

It is therefore a paradox that a vast number of audio systems have little or no difference between buttons for primary functions, such as 'play', 'stop' and 'next', and the buttons for the far less important functions such as menu, PTY, etc. Even the important volume function is often assigned to similar buttons or flat, hard to turn wheels.

The preset function is both liked and disliked. It is liked because of the ability to access one's favorite radio stations just by pressing a button, and disliked by many people who find storing these stations (to the given button) to be very difficult. There is something about the concept of presets that makes it hard to make good interfaces for storing that are also easy to understand and use. Structurally, one can store an instance – a copy of a radio station – to a button. How is it possible to design an interface that communicates this?

2.3.4   The What – Media structure

With just a number of sources, selecting both tracks and stations is currently done in a two-level selection hierarchy: 1 – source, 2 – track or station. A challenge for future systems is to encompass the large number of albums and the possibility of radio station groups. This could either be solved by introducing a third level in the selection hierarchy, 1 – source, 2 – album (and possibly group), 3 – track or station, or by letting the first selection step include a large number of sources. If the latter is possible, this is to be recommended since a third level in the selection hierarchy would add to the complexity of the interface.

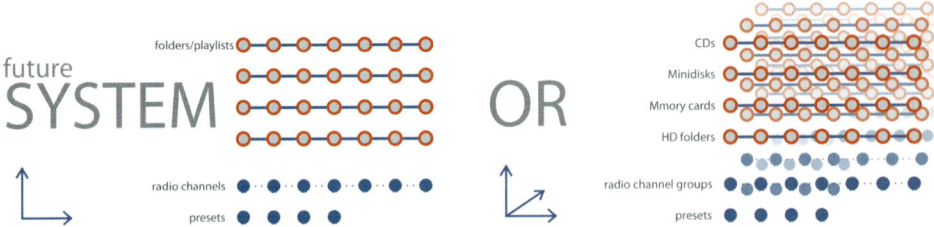

Figure 6.2 Future systems with 2- or 3-level hierarchy.

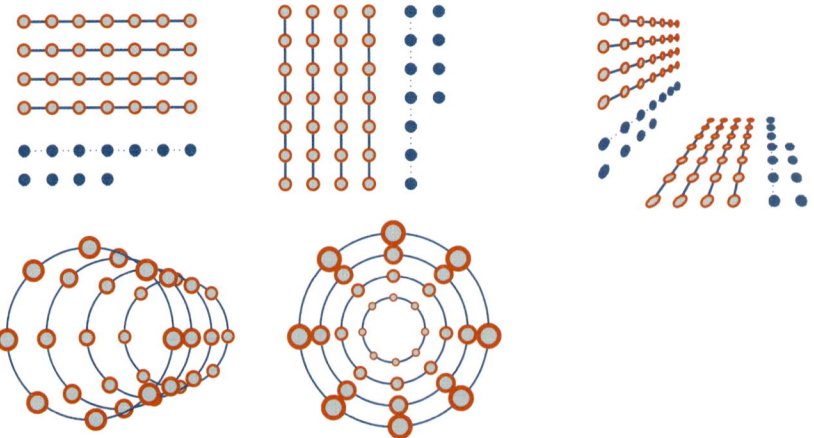

Figure 6.3 Different spatial directions of the media.

Another important question is the spatial directions of the different levels. As pointed out earlier, today's interfaces mainly work along a horizontal axis. We could therefore assume that users are customized to thinking about, for example, tracks as elements along a horizontal axis. However, the question is whether this the best direction to move in if we take the user into account. If, instead, we look at file structures in computer operation systems, which we can assume most users are familiar with, the smallest elements (files / tracks) are along a vertical axis.

2.3.5 Handling of radio

As mentioned earlier, storing and retrieving radio stations in presets is a popular but troublesome function. If a new interface is going to be based more on touch, how should this function be implemented?

There are three possible future approaches to presets that might be combined or chosen alone:

1. By displaying a (scrollable) list of radio stations, the user can find the desired station easily. This might eliminate the need for presets.
2. Technology such as force feedback makes it possible to identify the favorite radio stations haptically. Instead of copying the preset stations to buttons or other controls, the preset stations could be left among the other stations, but emphasized by a haptical signal. A list of radio stations could, for example, be felt like small tactile >ticks< until a preset stations appears and a larger >click< is felt. This >click< could also make it easier to stop at a preset than a not-preset station.
3. A solution similar to the current one could be developed, where preset stations are moved or copied to a new location.

Today, the user navigates through linear-scaled frequency bands, but there are areas in the bands that are much more densely populated by stations than others. A possibility is to use a non-linear scaled band. With RDS technology and the developing DAB/SDARS it is also possible, instead of frequency, simply to relate to the station name, possibly displayed in a list.

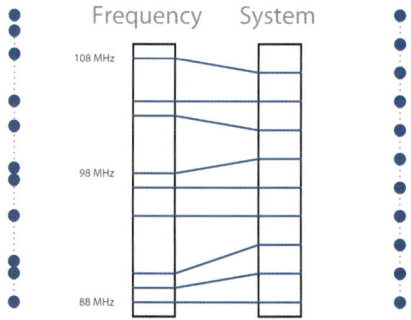

Figure 6.4 Radio stations could either relate to the position on the frequency band or be equally spaced along the system scale.

2.3.6 Handling of portable media

As observed in the preceding discussion in this paper, new types of movable media, such as minidisks, MP3 CDs and memory cards are already being used in car audio systems. Because of this fragmented media landscape and the relatively low costs of adding an extra input device, there are reasons to expect that future systems will let the user choose from a larger number of different sources than today. Hopefully, this will not mean a mass of different physical objects that will be loaded while driving, such as the change of CDs is today.

2.3.7 The How – Sound control

A user will need to modify the volume, but other properties in a high quality sound system, such as the Volvo system, rarely need to be adjusted. This means that the volume control should have a primary position and design. However, some users and some situations might make other sound control interesting and necessary. Balance and fader (the spatial position of the sound to the left/right and front/back) could, for example, be used if the sound is particularly interesting to some people in the car (e.g. children's programs for passengers in the back seats or news for those in the front seats).

Adjusting bass and treble is primarily related to audio systems of lower quality. It is possible to draw parallels to home hi-fi systems that do not have an equalizer. Some car audio systems also compensate for the noise from the road. Still, bass and treble might be a feature that some users or situations require.

The problem for most users arises if they want to adjust any of the above-mentioned sound controls (except volume). Since the different ways of controlling these do not give an apparent representation of what they in fact control, hence users are easily confused.

2.4 Constraints

Current audio systems and most navigational and communicational systems offer full functionality while driving. This is offered, even though the driver's workload and distraction is considered to be high, for example when he or she is entering a destination address into the navigational system.

It is possible to offer full functionality only while the car is stopped or driving at low speeds, leaving only the most important functions to the more demanding driving situation. An example of such a system is the Volvo television that turns off when accelerating beyond a certain speed. One of the challenges with such a solution is that even if it is in the best interests of the user, the user might be deeply annoyed by the forced limitations and search for ways of overriding the system.

General functions that could be limited to low speeds are demanding tasks, such as entering addresses into the navigational system, browsing the Internet and writing messages. In the audio system, complex functions which are rarely used could also be limited to low speeds. Examples of functions might be different RDS controls, bass, treble, storing and managing radio presets and playlists, and also having to physically load new media such as CDs. Limited functions could be inactivated, hidden, or constrained in a way that makes them impossible to use while driving. An example of a constraining operation could be that the ignition key is needed to operate the system or that the steering wheel needs to be folded down to reveal the control interface.

2.4.1 Placement of controls and displays

Many manufacturers in the car industry operate with a division between areas for driving related controls (e.g. around the steering wheel) and comfort-related functions (e.g. centre stack with climate and audio system).

Table 6.1 shows a summary of controls (haptic input to the system) that can be located at different places in the driver's vicinity. What is important is that they are easily reachable. The more the driver needs to reach out to use the controls, the more attention is directed to the operation and the more time it will take to get back to the normal driving position, which is significant if an unanticipated traffic event occurs. Some areas are particularly relevant for the audio system.

In the same way as for controls, displays (visual output from the system) can be located at different locations in the interior of the car (Table 6.2). It is essential to achieve good readability and to minimize eyes-off-the-road time. Volvo requires a display to be at a 30° downward view angle or above.

Table 6.1 Placement of controls.

Area	Possibilities	Limitations	Current use by Volvo and other manufacturers
Tunnel (area between the front seats)	Easily reachable. Possibility to support hand while operating control.	Limited to tactile feedback, since it departs heavily from the normal line of sight.	Usually used for armrest and/or small compartments for various small items such as keys, CDs and receipts. Cup holder.
Mid-console	Good visibility. Reachable.	No support for the hand.	Used as a general placement for controls for systems such as climate and audio.
Steering wheel	Good visibility. Easily reachable with both hands.	Controls might interfere with the operation of primary controls. Excludes other passengers.	Some use for basic controls for cruise control and audio system.
Door	Visible. Reachable.	Left-hand operation. Line of sight departs from the normal line of sight.	General "room controls" such as window, mirror and central locking.

Table 6.2 Placement of displays.

Area	Possibilities	Limitations	Current use by Volvo and other manufacturers
Mid-console	Can be located together with controls.	Might be limited in space.	Used for smaller displays such as audio system and climate.
Above mid-console	Better visibility.	Cannot be located together with controls.	Used as output unit for navigational systems.
Steering wheel	Good visibility. Can be located together with controls.	Might be mistaken for primary display behind steering wheel. Airbag operation. Excludes other passengers.	Presently not utilized.
Instrument panel	Good visibility.	Cannot be located together with controls. Might interfere with primary display.	Mostly limited to primary display as vehicle information and warning signs.
Windscreen / above steering wheel	Good visibility.	Cannot be located together with controls. Will obstruct view through windscreen.	Some rare examples with simple head-up displays.

As shown in Table 6.2, co-localization of both control and display is only offered in the mid-console and on the steering wheel. An alternative would be to mount the control / display on a more or less flexible arm that was both easily reachable and offered good visibility.

2.4.2 The sacred cup holder

Although this might sound odd, the cup holder seems to be more or less sacred in the interior of a car. This is especially true for the North American market, where a functional, centrally placed cup holder is important for acceptance and customer satisfaction. Some car models have cup holders in the centre stack while others (including Volvo) have them in the tunnel behind the gear lever. This reduces the space available for controls. When BMW implemented the iDrive in the BMW 7 series they solved this problem by moving the gear lever to the space behind the steering wheel.

2.4.3 Technological possibilities

Healthy conservatism has until now kept most of the surge for technology away from the driver's environment. Although available, much of the technology has been seen as immature and poorly adapted to the driver. Whether the "techies" have grown impatient or the technology has matured is unknown, but the technology has certainly been taken in from the cold. It seems as if a screen-based menu interface will become a common element in tomorrow's top-end cars. BMW's iDrive is an extreme example of this. Here, the driver is supposed to control over 700 functions by navigating a menu structure with a joystick control. More than 100 functions can also be voice controlled. The challenge is not to be excessive, but to utilize technological possibilities where and when they are useful. While some technologies are already used in in-car interfaces, others need a longer time of development for them to become adapted to the application. The feasibility of such technologies depends on the project's budget and time perspective.

2.4.3.1 Input

Voice recognition is a new system input that has become economical and technologically feasible in recent years. The BMW iDrive has around 100 functions, which could be voice operated. JVC and Sony sell car stereos (320–400 Euro) with simple voice control.

More experimental is gaze and gesture recognition that is under development for various applications and systems. This technology makes it possible to recognize where the driver looks and what movements he or she makes.

Although visual distraction should be kept to a minimum, touch screens could be a usable technology if display screens are unavoidable in the development of interior car design. These give direct control of elements in the screen interface.

2.4.3.2 Output

There is a general trend toward bigger and better screens in cars. These are shared between different systems, such as the audio, climate and navigational systems. As the technology advances the prices go down and resolution and colors improve.

Apart from the development towards bigger and better screens in the driver's environment, there are other system outputs that one might explore. There are big developments in the field of interior lighting in the car industry today. Illumination is mainly explored to enhance the feeling of roominess and quality. However, this could also be used to give forms of output from interfaces in the car.

2.4.3.3 Haptic feedback

Traditionally, communication via interface devices has been a one-way progress. One presses a button or moves one's mouse, joystick or wheel, and the device communicates a signal to the system. Feedback comes from a different device such as an LED, display or speakers, or as a system response.

Tactile feedback adds technology to control devices that allows them to communicate to the user as well. Today, the two primary types of tactile feedback for control devices are vibration feedback and force feedback.

Force feedback is a sophisticated technology and provides people with very realistic tactile feedback. It has been widely used for some time in medical, space and flight simulators to provide life-like training for students and professionals who make split-second decisions based not just on sight and sound, but also on their sense of touch. Force feedback devices include one or more motors plus additional mechanical and electrical components and are capable of a wide variety of effects. Force feedback can either resist the operator's movement or assist it. In addition, it also provides specific directional movement and information.

Vibration feedback works on the same principles as the vibration mode in a mobile phone or pager, but typically is much more sophisticated. It can be used to communicate specific ideas or sensations. The effects can range from a simple buzz that serves as an alert or confirmation, to a slower rhythmic motion that simulates vibrations from an approaching train, to a single, strong jolt that simulates a collision. However, vibration feedback cannot resist the user's movement, assist it, or provide any directional movement or information. Because they are much simpler, vibration feedback devices tend to be much less expensive than force feedback devices.

2.4.3.4 Extending the interface – Ambient display

For the audio system, the playback of sound is, of course, a very direct form of feedback that tells the user a lot about the system, such as whether it is playing, what it is playing, and how the sound is adjusted. However, the auditive output could also be refined to give more feedback-like information about track and source changes by utilizing dynamic modulation of the sound, particular sounds (earcons) or the capacity of a surround system.

Although modern radios do not send out noise while tuning, a noise similar to the tuning noise made by older models could help the user to understand the operations of the system. Another example of this could be the sound that a CD makes when it starts spinning. Such simulated noises are different from the ones used in the Microsoft Windows operating system because they are more subtle and integrated into the overall sound.

The Tangible Media Group at MIT Media Laboratory defines ambient displays as presentations of information within a space through subtle changes in light, sound and movement, which can be processed in the background of awareness. The use of ambient displays is a new approach to interfacing people with digital information that does not give exact information but rather an understanding of the general state of a system. In the car, this could be used by making use of interior lights, (surround) sound or a light movement of air on the skin.

3 Design

In the early design phase it is important to pursue parallel design concepts in order to avoid mental preoccupations, as well as to utilize ideas across concepts. As this is a central feature of the design process, we present the four most promising concepts from the early design phase. The concepts emerged from one workshop held in the Department of Product Design, at the Norwegian University of Science and Technology, in Trondheim, and one workshop at a major Swedish/American car manufacturer's headquarters in Gothenburg. We present here only the main interface, not the secondary interface (e.g. bass, treble, etc.).

3.1 The four most promising concepts from the early design phase
3.1.1 The centre stack interface

Developing an interface positioned in the centre stack is the most conservative solution when it comes to industry standards and user experience. There is much that could be explored and possibly improved:

- Division between primary and secondary functions – The most important controls could be made bigger and placed more centrally than the ones seldom used.
- Hiding functions – The controls that are seldom used could be hidden while driving, either physically or by turning off illumination, etc.
- Tactile clues – It is fairly easy to give the user more tactile feedback and hence less visual distraction. Important buttons could stand out and different areas could have different textures and clearly indicated borders. This would also improve visual contrast.
- Structural improvement – It is possible that a sequential layout of the controls could make the operation clearer to the user. It could, for example, be natural to organize main controls in the following sequence: 1. Source; 2. Track/station; 3. Volume.

3.1.2 The gear level analogy

A very good example of a touch-based interface in the car is the gear lever. There are hardly ever situations were the driver needs to look down at the gear lever to know how to operate it or what position it is in. This is simply sensed with the hand/arm and by listening to the sound of the engine. It would be interesting to make use of this quality in an interface for the audio system.

A mechanical arm, somewhat similar to the gear lever or a smaller slider could control the source and/or track. Similar to changing gears, one could change to different audio sources. Such an interface would probably be best positioned in the tunnel, behind the gear lever.

The analogy with the gear lever should not be drawn too far. A small lever positioned close to the gear lever could very well be mistaken for the 4wd or low gear. The user might also have wrong expectations from using a large mechanical

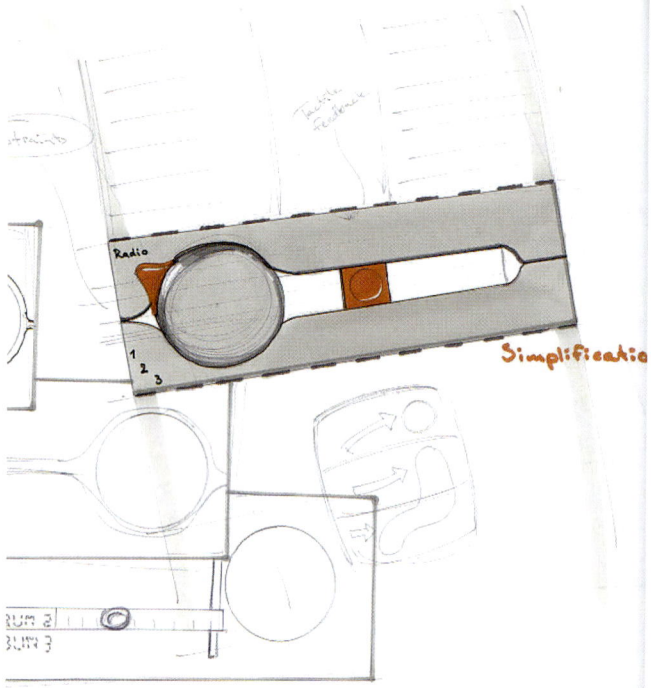

Figure 6.5 Sketch of center stack solution.

handle, expecting it to control something mechanical, such as something that has to do with the motor or transmission. In contrast, the audio system is very non-mechanic. Hence, it could be interesting to explore the possibility of introducing a smaller "amulet", reflecting this nature of the audio system.

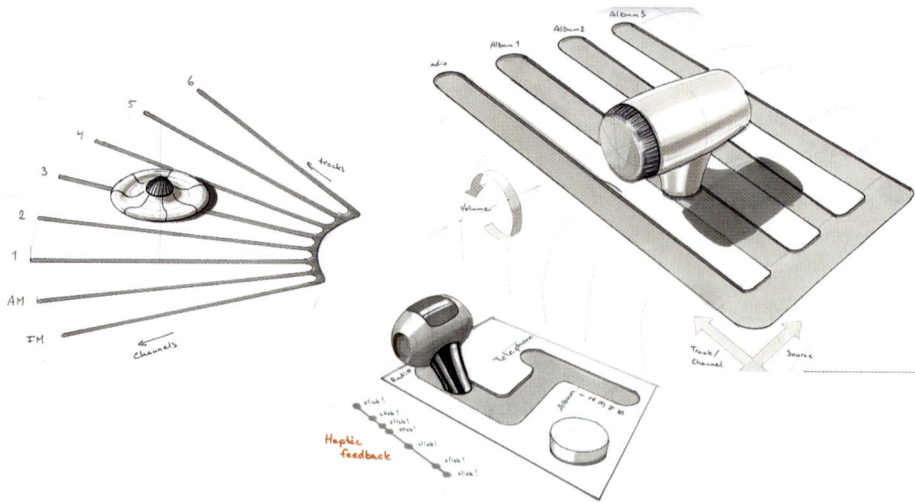

Figure 6.6 Sketch of gear level analogy.

3.1.3 The rolling cylinder

A pleasant interface for going through many elements is the roll. Various solutions utilizing a rolling cylinder have been tried out during this project. The most natural solution may be to roll through tracks or stations, while *source* is chosen with a different control. To make use of the haptical qualities of the hand, the roll should be of some size, but at the same time making it pleasant to grip. This also opens up the possibilities of allowing such an interface be located where the hand can rest, leaving the area behind the gear lever in the tunnel as a natural position for the interface.

The volume could be controlled by a similar concentric roll. It is important then, to clearly mark the difference between the two rolls, since there are no structural analogies between the two controls. Still, aesthetically and technically, such a solution would be neat and tidy.

To keep focus on tactile qualities, a slider could control the source change. The handle on the slider would metaphorically become "a play head", reading the information on the roll. It is then possible to picture the roll as a stack of discs that lie horizontally. When sliding the play head along the pile's axis the active disc is changed and when rolling the pile, the play head reads different tracks within the disc.

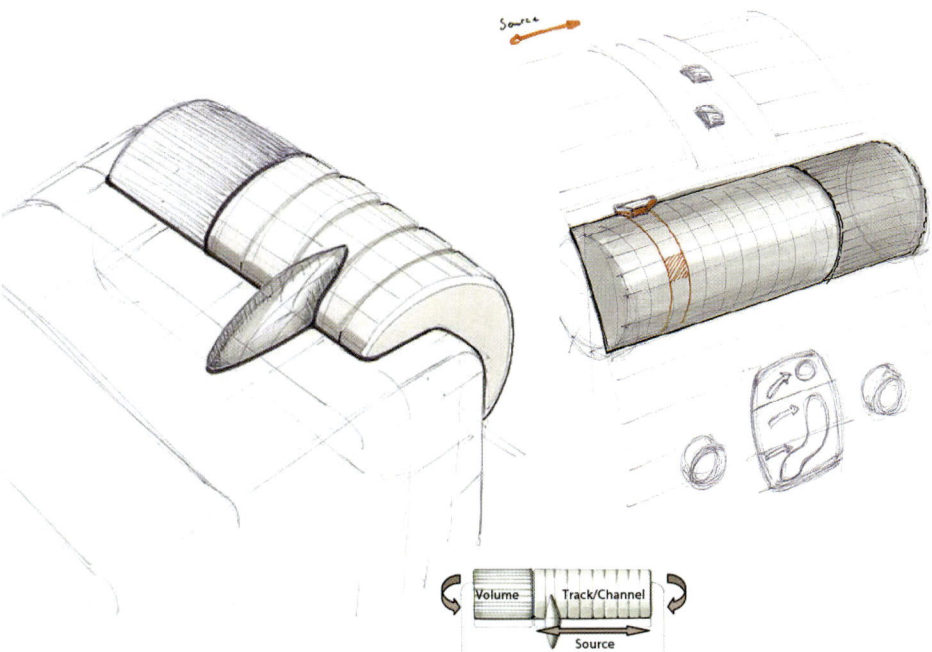

Figure 6.7 Sketch of the rolling cylinder

Embodied Minds – Technical Environments

3.1.4 The gesture-based interface

To keep the mind open to untraditional ideas, a more conceptual idea has been worked on. Based on more-or-less advanced technology it is possible to register the user's hand gestures. This could prove to be a natural way of controlling the system. What if, for example, one raised the volume by waving one's hand upward and changed track by twisting it clockwise? The problem with such interfaces is that often they are intended to be natural, but end up being a range of commands that the user needs to learn. The reason for this is that they do not have any physical affordances that invite the user to use the accepted commands.

MIT Medialab is one of the institutions working on what some call "phicons". Phicons are physical objects that work as tools or representations of digital elements. An often-used example is the "marble answering machine", where a marble rolls out of the machine every time there is a new message. If one drops a marble into the machine, the message represented by the marble is played. A problem with such solutions in a car is the number of loose objects.

One solution is to combine a physical and non-physical interface. Physical objects or forms could function as abstract "containers" for music, while the music is played in an "active area" where the playing is controlled and manipulated. However, instead of picking up the containers, the user simply picks up air from the container and releases it onto the active area. This could easily be censored by, for example, inductive sensors in the surface of the different physical elements.

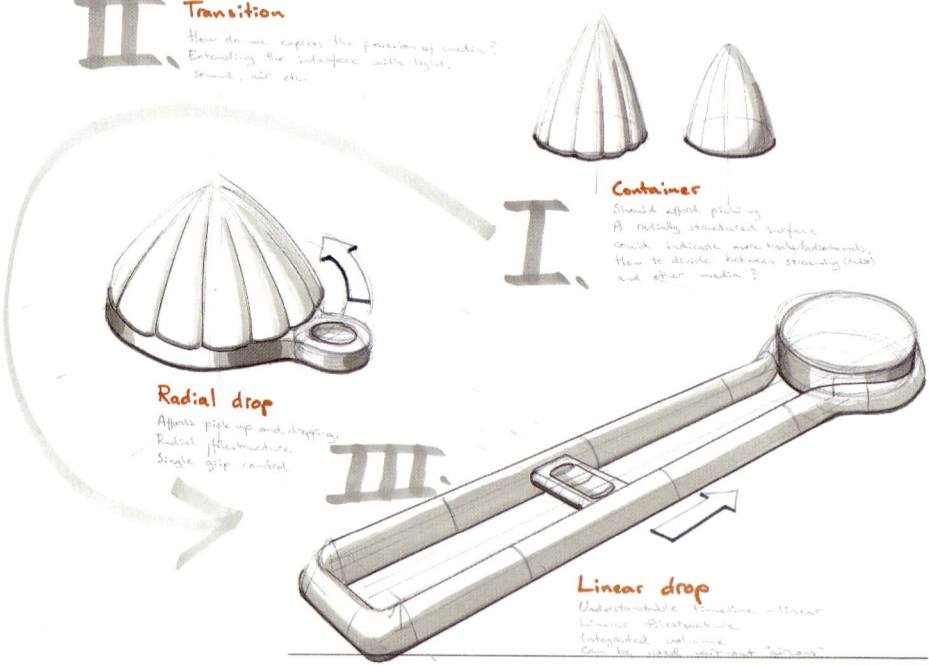

Figure 6.8 Sketch of the gesture-based interface

3.2 Choice of concept

The different concepts for the main interface all have some good qualities, but common to them all is that while they might look good and interesting on the drawings and in the explanations they still need to be worked out in detailed and tested before they can be evaluated fully. Unfortunately, in this project, there was no time to develop all of them, so it was necessary to make a selection. When categorizing different solutions, both the technical feasibility in development and production, and the possibilities of making good prototypes that could be evaluated within the limits of this project were evaluated.

Developing an interface in the centre stack would in many ways be the most straightforward choice. However, to enable this project take a new look at interfaces in cars and to keep a clear focus on non-visual information this option was eliminated. The choice then fell to the roll concept. This was largely because of the way this would effectively solve the challenge of large lists of tracks / stations as well as the potential of the roll to comply with the EIP properties described above.

Table 6.3 Evaluation of concepts.

	Placement	Appeal	Usability challenges	Technical feasibility in development and production	Prototype feasibility within this project
Center stack	Center stack	Accepted placement. Everything in one place.	Clean up and make a simple interface. No support for the hand.	Very adapted to current practice.	Digital display and lights are the biggest challenge. Could be done in 2D on a computer.
Gear lever	Tunnel, supporting interface in center stack and/or door.	Good qualities as gear lever. Explicit focus on tactility.	Must not become a copy of iDrive or the gear lever. It should be apparent what it can and can not do.	Mechanical parts are a challenge.	Difficult to make a prototype that feels, looks and works right.
Roll	Tunnel or center stack. Possible supporting interface in door.	Roll invites interaction. Evident linear direction. Explicit focus on tactility.	Unintentional activation.	Relatively small and compact.	Should be possible to make prototype. The feel is the biggest challenge.
Gesture	Anywhere.	Highly novel. Would re-define "phicons".	The user needs to understand it …	A challenge, but should be possible.	Might possibly be simulated (Wizard of Oz, etc).

3.3 Presentation of prototype

The functions identified as the most frequently used functions in in-car audio systems are the on/off, volume, source selector, and navigation through the source (track/station). These functions are controlled by the roll interface placed between the gear lever and the arm rest.

In Figure 6.9, the interface consists of two cylinders, a slider in front of the cylinders and a button on the left-hand side. The smaller of the two cylinders, to the left, controls the volume at the same time as it turns the system on and off when the volume is turned down completely. The slider in front slides along the largest

Figure 6.9 Rendering / photo manipulation of the physical interface, shown in a Volvo S80.

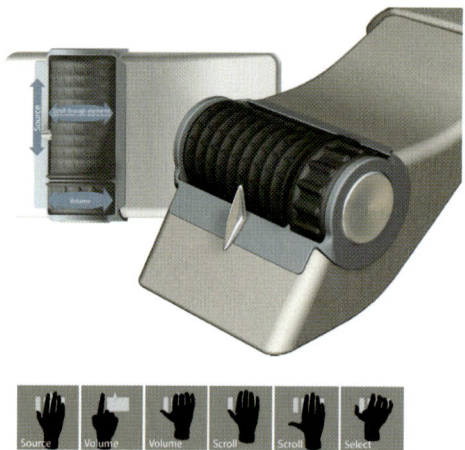

Figure 6.10 Main functions and interaction concept.

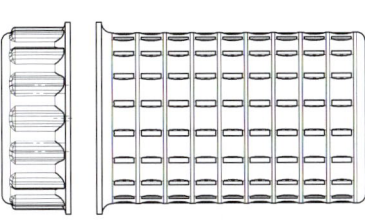 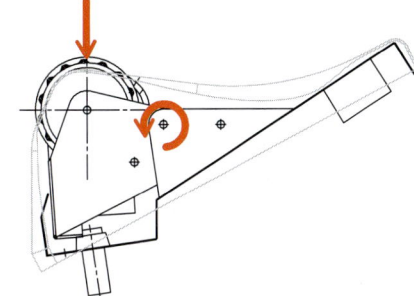

Figure 6.11 The volume roll: a) grooves and dents on the two cylinders, and b) the selection mechanism.

cylinder and changes the current source (radio, phone, CD 1, CD 2, etc.). The largest cylinder, to the right scrolls through the chosen source. The button on the left is used for selecting, for example, phone numbers from the phone list or radio stations to store as presets.

To illustrate the structure of media in horizontal sources and vertical tracks/stations/list elements, the main roll was designed to reflect this. Main grooves are spaced along the source slider, while smaller dents mark the smaller elements within these. It is hoped that the groove along the source selector would emphasize the play head metaphor and the choice of sources.

The volume roll has only one dimension and should reflect this in its form. To make a good thumb-grip, the volume roll was designed with large knobs. To reduce the number of buttons on the physical interface in the tunnel it was decided to turn off the system by turning the volume down until a click could be felt. The source selector was designed to have a low profile, so as not to get in the way of the hand when operating the rolls.

3.3.1 Haptics

The cylinders and the source selector can have either conventional, purely mechanical haptic feedback, or more advanced mechatronical haptic feedback:
- A mechanical haptic feedback such as the one used in the prototype is typically embedded in the rotary sensors of the cylinders and in the sliding mechanism of the source selector.
- A mechatronical haptic feedback requires electrical motors and motor controllers and offers the ability to dynamically alter haptic feedback, dependent on the current source or position. The interface could divide between ordinary radio stations and presets. Also, stored media could fast forward or fast rewind if the cylinder is turned slightly.

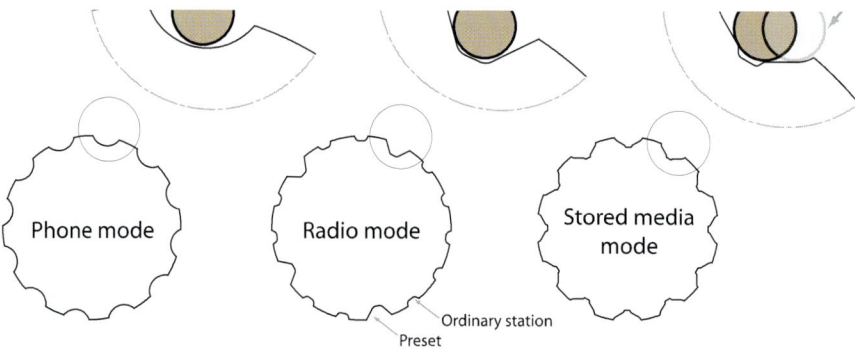

Figure 6.12 Mechatronical haptic feedback allowing different haptical patterns for phone, radio and stored media, shown here with a mechanical analogy.

3.3.2 Construction

Figure 6.13 Construction.

Dependent on the choice of technology for the haptic feedback, the two cylinders need to be connected to either 1) rotary sensors with haptic increments or 2) electromotors with motor controllers censoring the position of the cylinder and controlling the motor's feedback. The construction is compact and hence it is a realistic option to apply the interface in various interior designs.

If possible, the cylinders should have some weight, in order to give the right feeling of quality and precision. To prevent the fingers from slipping and to give a positive experience they should be covered with a layer of soft, high-friction material.

3.3.3 Screen interface

The screen interface has been developed based on two different scenarios:
- A conservative scenario where the future sound system uses a relatively small, single colour screen, like the ones used today positioned in the centre stack. However, there are reasons to expect that the screen would have higher resolution.
- A more high-end scenario where a large colour screen will be shared between different systems, such as the climate system and sound system. The position will then most likely be high up in the centre stack.

The screen is intended to support the interaction with the system and to give confirmation whenever the user wants it.

3.3.4 Smaller screen scenario

The concept for the small screen scenario is focusing on simplicity, but yet taking advantage of the higher resolution. It is proposed that a pictogram to the left would clearly show the user which source is active.

Rapid transitions between different sources and elements (station, track, phone number) cause the screen to link with the physical interface. The transitions move in the same direction as the corresponding physical action.

Figure 6.14 Screen interface on the small screen.

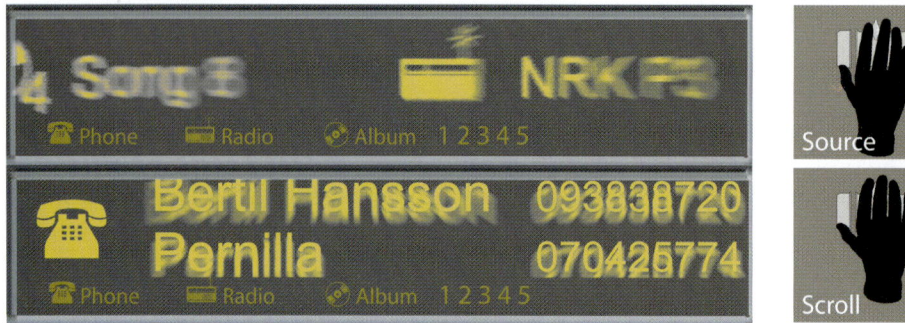

Figure 6.15 Transitions (from right to left and up).

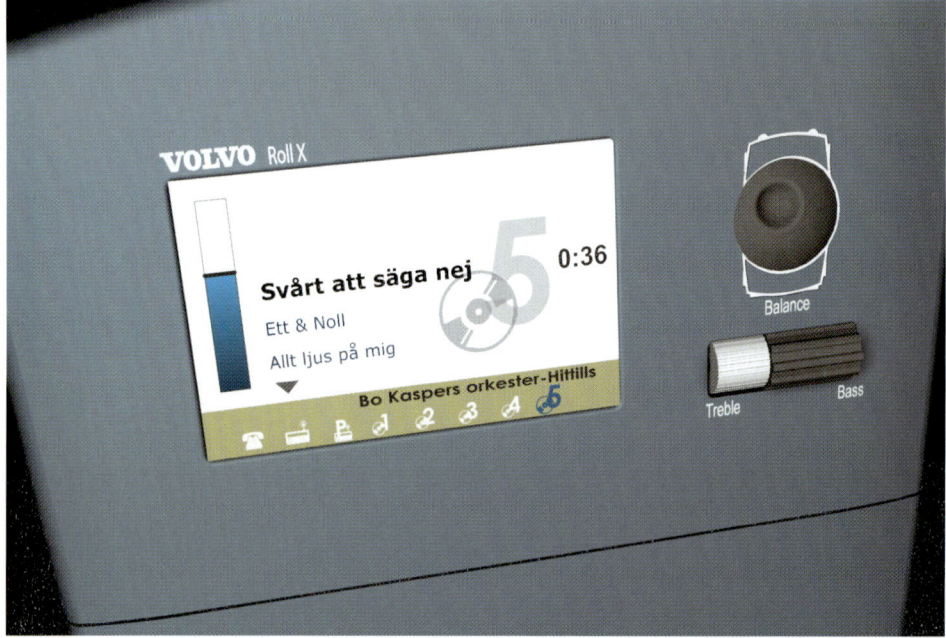

Figure 6.16 A large screen.

In addition, the concept based on the large screen scenario focuses on simplicity, but aims at utilizing the advantages of a bigger screen. The graphical liberty allows the screen interface to correspond with the physical interface. It is, as the physical interface made up by three main areas: volume to the left, element (phone number, channel or track) to the right, and source at the bottom. It is expected that the use of sound formats such as MP3 will make it possible to display song titles and album titles.

This concept also uses pictograms to clearly show the user which source is active, both at the bottom and in the background of the main window. Not only the position and sequence, but also the transitions of the source area aim at giving a good parallel to the source selector on the physical interface.

Figure 6.17 Source pictograms.

Figure 6.18 Presets.

With a larger screen it is possible to show not only the current playing element but also the previous and next elements in the list. This makes it easier and faster to find the desired element. An arrow at the bottom and top of the list indicates whether there are more elements below or above the list.

With such easy navigation through available radio channels, there might not be any need for preset functionality. Still, an alternative is to let the user store their favourite radio channels in a separate source containing a shorter list of only favourite channels.

3.3.5 Form follows function

With an additional screen, the user relates to three different parts of the total interface, with three different sensory modalities: tactile, auditive and visual. To use this as an advantage, the three modalities need to be tightly coupled and often make each other redundant. By doing this, the user becomes less dependent on each modality.

Technically, there is an electronic system connecting the different interfaces, interpreting the signals and broadcasting sound to the speakers. When using the system, the user will form a mental image of the structure and working of the system. It is important that the different interfaces communicate a mutual internal structure.

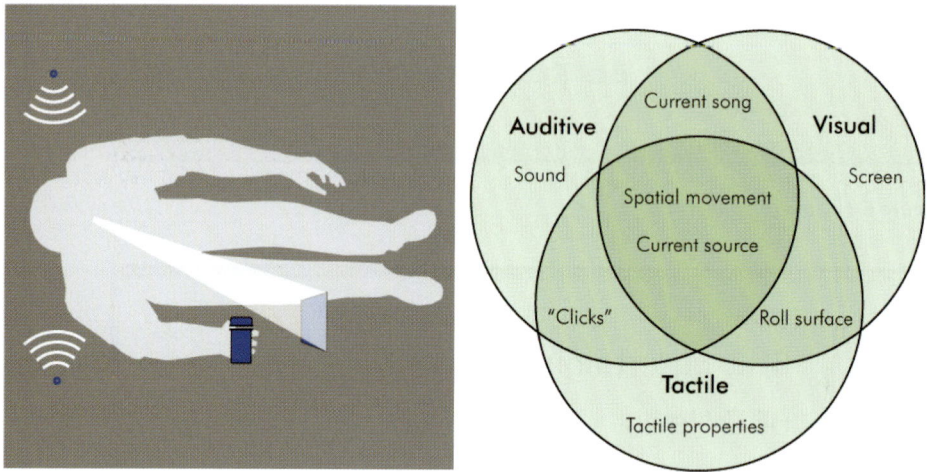

Figure 6.19 Different elements of the interface need to communicate a mutual structure.

3.3.6 Design details

The main roll was designed to reflect the structure of media in horizontal sources and vertical tracks/stations/list elements. Main grooves are spaced along the source slider, while smaller dents mark the smaller elements within these. It is possible that the groove along the source selector would emphasize the play head metaphor and the choice of sources.

The volume roll has only one dimension and should reflect this in its form. To make a good thumb-grip, the volume roll was designed with large knobs. To reduce the number of buttons on the physical interface in the tunnel it was decided to turn off the system by turning the volume down until a click could be felt.

The source selector was designed to have a low profile, not getting in the way of the hand when operating the rolls.

A *select* function is needed, for example, to let the user store radio stations in presets. On the prototype, this was done by pressing on the cylinders.

Different screen layouts were tested and evaluated. Common for them all was that they used the same physical structure as the physical interface with source horizon-

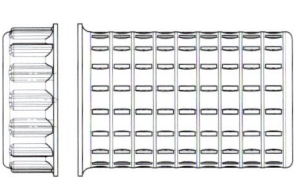 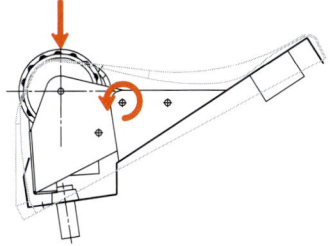

Figure 6.20 The volume roll: a) grooves and dents on the two cylinders, and b) the selection mechanism.

Figure 6.21 Concept for the screen interface.

tally placed at the bottom, volume to the left, and the list elements to the right. This was chosen to make the layout more understandable, and at the same time it proved to be a good layout.

4 Discussion

Since the introduction of the in-car radio in the early 1920s, very little progress has been made with respect to the user interface. Although there are no longer any technological constraints that strongly shape the design solution (with digitalization there are no constraints between the representation and the represented) and, in principle, the interface looks like it did 50–60 years ago. The phenomenon that interfaces develop in small incremental steps, even though there is no technological reason for this, can be termed *evolutionary design*.

In some cases, there are motives governed by economy behind the evolutionary design process; that is, the marketing of a product or a product class depends on familiarity with other products, or earlier versions of the product. However, in terms of effectivity, safety and enjoyability, there are very few good arguments for pursuing evolutionary design.

The most important causes for the ineffective, restrictive and conservative evolutionary design process are psychological. The term cognitive economy refers to the fact that the cognitive system will economize the processing of information wherever possible. The fact that people, when making an estimation, tend to lie close to an already established estimation, even though there is no rational reason why this estimate should be any better than any other, is termed the anchoring effect (Plous,

1989). Hence, it is cognitively economical to start off with previous products or similar products as a point of reference, even though the inherent weaknesses of these are almost certain to be transferred to the new product.

At an even deeper level, the problem seems to be that people tend to confuse intension with function. That is, people tend to think of the task in terms of the *thing*, instead of the task itself. For humans, it is very difficult to separate between intension and function, even though in reality there is only an arbitrary relation between the two. It is of paramount importance that designers are aware of this mechanism, and that proper methodology is used in order to work around the problem. For complex sociotechnical systems, Vicente (1999) has developed a device-independent design methodology. Vicente (1999) also considers Beyer and Holtzblatt's (1998) 'Contextual Design' methodology as device independent, because of the heavy focus on context apart from the technology as such. For smaller products with less complexity, there are some useful techniques which defy the intension/function problem.

One example is the forced functions technique, where the functionality of one artifact (e.g. a stapler) is mapped onto the product one is designing (e.g. a mobile telephone). Another example is user workshops, where users are asked to use devices which have no function (i.e. foam blocks or pieces of wood) as a means of 'conducting' a task. Other techniques which implicitly deal with the intension/function problem are those where one alters perception/action qualities of imagined users (remove, add or alter senses, limbs, gravity, etc.), or create completely altered worlds (Sci-Fi is often used as a source of inspiration). These methods were probably not developed in order to tackle the intension/function problem, but they are very helpful nonetheless, and should be used if the aim is to develop safer and more humane technology.

The aim of this paper has been to demonstrate that in addition to performing careful analysis of the activity of use in a particular context (and not paying too much attention to current technology and current interface solutions), one needs an explicit account of the properties of the human / technology system which are absolute, and not dependent either on the user, task or context. Classical cognitive psychology failed to develop such an account, but that does not mean that the questions they asked were ill-formed. On the contrary, the vision that technology should be an extension of human abilities and to augment the mental properties of humans in the same manner as an excavator augments the physical power of man still has a definite appeal. Theoretical investigations of how such a task might be pursued, e.g. the constraints that have to be taken into consideration, has been put forward by several authors (e.g. Flach et al., 1995).

Up to this point, however, very little empirical research has been done to uncover the general rules that invariably describe how different system representations give rise to different user experiences. The case presented in this paper demonstrates that even the scarce amount of empirical research conducted this far can have a significant impact on the final design solutions, and that the effectivity, safety and enjoyability of these products have the capability of outperforming products based on evolutionary design and general heuristics.

The aim of this paper has been to describe the theoretical background of the interaction design problem, and to document a design case that deals explicitly with these problems. No attempt has been made to empirically reveal whether the interaction concept presented here outperforms traditional interfaces in terms of workload, driving performance, etc., or at theoretically analyzing the qualities of the interface in terms of the EIP properties described in the introduction. The former is beyond the scope of this paper, while the latter is prone to subjectivity on behalf of the authors. However, the reader is encouraged to compare the qualities of standard in-car audio system interfaces and the interaction concept presented here with respect to the EIP properties.

References

Card, Moran & Newell (1983). *The psychology of human-computer interaction*. Hillsdale, NJ: Lawrence Erlbaum.

Cesari, P. & Newell, K. M. (2000). Body-Scaled Transitions in Human Grip Configurations. *Journal of Experimental Psychology: Human Perception and Performance*, 26, 1657-1668.

Flach, P.A. Hancock, P. Caird, J. & Vicente, K. (1995). *Global Perspectives on the Ecology of Human-Machine Systems*. NJ: LEA.

Gardner, H. (1985). *How mind's new science: A history of the cognitive revolution*. Basic Books.

Gibson, J.J. (1979). The Ecological Approach to Visual Perception. NJ: LEA.

Gibson, J. J. (1966). *The Senses Considered as Perceptual Systems*. Boston: Hougthon Mifflin Company.

Hancock, P.A. & Chignell, M.H. (1995). On Human Factors. In: Flach, Hancock, Caird and Vicente (Eds.) *Global Perspectives on the Ecology of Human-Machine Systems*. NJ: LEA.

Hoff, T. (2004). Comments on the Ecology of Representations in Computerised Systems. *Theoretical Issues in Ergonomics Science*, 5, 5, 453-472.

Hoff, T., Øritsland, T.A & Bjørkli, C.A. (2002a). Ecological Interaction Properties. In: P. Boelskifte, J. B. Sigurjonsson (Eds): *Proceedings of NordDesign 2002*, p. 137-151.

Hoff, T., Øritsland, T.A & Bjørkli, C.A. (2002b). Exploring the Embodied-Mind Approach to User Experience. In: Proceedings of NordiCHI 2002, p. 271-274. New York: ACM Press.

Hoff, T. & Hauser, A. (2008). Applying the Ecological Approach to Interface Design on Energy Management Systems: Developing a Compact System State Display (CSSD). *PsychNology*. In press/Accepted

Hoff, T. and Øvergård, K.I. (2008/This issue). Ecological Interaction Properties. In: Hoff, T. and Bjørkli, C.A. (Eds.): *Embodied Minds – Technical Environments: Conceptual tools for analysis, design and training*. Tapir Academic Press

Hofsted, C. (1996). Development of prospective control as the basis of action development, *Infant Behavior and Development*, 19, Supplement 1, 32.

Kobe, G. (2000). Death by Distraction. *Automotive Industries*, p. 30-39.

Kosslyn, S. M., Ball, T. M. & Reiser, B. J. (1978). Visual images preserve metric spatial information: Evidence from studies of image scanning. *Journal of Experimental Psychology: Human Perception and Performance*, 4, 47-60.

Lakoff, G. & Johnson, M. (1999). *Philosophy in the Flesh: The Embodied Mind and its Challenge to Western Thought*. New York, NY: Basic Books.

Lee, D. N., Davies, M.N.O., Green, P.R. & Van Der Weel, R. (1993). Visual control of velocity of approach by pigeons when landing. *Journal of Experimental Biology*, Vol 180, Issue 1 85-104.

Nardi, B.A. (1996). *Context and Consciousness. Activity Theory and Human-Computer Interaction*. London: MIT Press.

Norman, D. A. (2002). *The Design of Everyday Things*. New York: Basic Books.

NHTSA (2001). *Traffic Safety Facts 2001*, U.S. Department of Transportation National Highway Traffic Safety Administration, DOT HS 809 484

Plous, S. (1989). Thinking the unthinkable: The effect of anchoring on likelihood estimates of nuclear war. *Journal of Applied Social Psychology*, 19, 67-91.

Ranney, T., Mazzae, E., Garrott, R., & Goodman, M.J. (2000). *Nhtsa driver distraction research: past, present and future*. [On-line Paper]. http://www-nrd.nhtsa.dot.gov/departments/nrd-13/driver-distraction/Welcome.htm

Rasmussen, J. (1983). Skills, rules and knowledge; signals, signs, and symbols, and other distinctions in human performance. IEEE *Transactions on Systems, Man, and Cybernetics*, SMC-13, 257-266.

Santana, M. & Carello, C. (1999). Perceiving whole and partial extents of small objects by dynamic touch. *Ecological Psychology*, 11(4): 283-307.

Sternberg, R.J. (2008). Cognitive Psychology 5th Edition. Belmont, CA: Wadsworth Publishing

Stutts, J.C. (2001). *The Role of Driver Distraction in Traffic Crashes*, prepared for AAA Foundation for Traffic Safety by University of North Carolina Highway Safety Research 2001

Turvey, M. T., Shaw, R. E., Reed, E. S. & Mace, W. M. (1981). Ecological laws of perceiving and acting: In reply to Fodor and Pylyshyn (1981). *Cognition*, 9, 237-304.

Vicente, K.J. (1999). *Cognitive Work Analysis – Toward Safe, Productive, and Healthy Computer-Based Work*. LEA.

Vicente, K. J. & Rasmussen, J. (1992). Ecological interface design: Theoretical foundations. *IEEE Transactions on Systems, Man and Cybernetics*, 22(4), 589-606.

Warren, W. H. Jr. (1995). Constructing an Econiche. In J. Flach, P. Hancock, J. Caird & K. Vicente, (Eds.), *Global Perspectives on the Ecology of Human-Machine Systems*. (pp. 210-237), Hillsdale, NJ: Lawrence Erlbaum Associates.

Warren, W. H. Jr. (1984). Perceiving affordances: Visual guidance of stair climbing. *Journal of Experimental Psychology: Human Perception and Performance*, 10, 683-703.

Warren, W. H. Jr. & Whang, S. (1987). Visual guidance of walking through apertures: Body-Scaled information for affordances. *Journal of Experimental Psychology: Human Perception and Performance*, 13, 371-383.

Woods, D.D. & Watts, J.C. (1997). How Not to Have to Navigate Through Too Many Displays. In Helander, M.G., Landauer, T.K. and Prabhu, P. (Eds.) *Handbook of Human-Computer Interaction*, 2nd edition. Amsterdam, The Netherlands: Elsevier Science.

Zaff, B. S. (1995). Designing with Affordances in Mind. In J. Flach, P. Hancock, J. Caird & K. Vicente, (Eds.), *Global Perspectives on the Ecology of Human-Machine Systems*. (pp. 238-272), Hillsdale, NJ: Lawrence Erlbaum Associates.

Part III:
Simulator Skills Training

# 7

The Role of Fidelity, Transfer and Cognitive Involvement in Learning
– A review of simulator training

Paul Andreas Lundeby

This paper explores the role of simulator training in relation to different levels of human cognitive functioning, pointing out the importance of having a differentiated view of such training, and that a reciprocal relationship exists between levels of cognitive functioning and different kinds of simulators. This is carried out in part by using the SRK taxonomy by Rasmussen and by discussing the relationship between fidelity and transfer in order to gain a broader and deeper understanding of this relationship. Finally, some principles of design in relation to functional simulator training are presented.

1 Introduction

In recent decades the belief in the potential of gaming and the use of simulators[1] to facilitate learning has grown immensely (Ruben, 1999). The gaming industry has, since its early start in the mid-1970s, grown to be multi-billion dollar business, with a revenue of USD 31 billion worldwide in 2003 (Wikipedia.com). Financial growth is often accompanied by development and this development has contributed to making games and simulations into the solution to a wide range of learning objectives. However, to date, few discussions have focussed on the fundamental issue of differentiating between different kinds of simulators in relation to different levels of

---

[1] A simulation is an imitation of some real thing, state of affairs, or process. The act of simulating something generally entails representing certain key characteristics or behaviours of a selected physical or abstract system (www.wikipedia.com).

human functioning. Discussions up until now have appeared polarized and narrow, treating simulators as a fixed concept or as a general category of learning resources rather than a dynamic framework which requires customization depending on the area of use. In addition, the discussion has avoided addressing the fit between learning methods and skills acquisition. It is important to identify the skills in question before starting to address different kinds of simulators. Skills vary in complexity and degree of cognitive involvement. They range from simple motor movements and other routine tasks in everyday activities to high-level intellectual skills. It is claimed in this review paper that a discussion without such a distinction is questionable for several reasons. First, simulators for one set of skills, such as negotiation, leadership or decision making, cannot be seen simply as an adjustment of simulators for procedural skills, such as flight simulators, treating the differences between the two as simply content dependent. A simulator for negotiation would have to rely on different characteristics of the user in order to enhance learning, such as strategic thinking, reasoning, planning, communication, and reflection. Second, the degree of fidelity[2] also differs greatly between the different types of simulators. Traditionally, flight simulators, and the like, have had a high degree of apparent fidelity. However, this is not the case in many simulators associated with the training of more subtle and implicit skills.

Furthermore, games and simulations can only be as effective as the pedagogical approach that is the basis of their design and development. This demands a thorough understanding of the processes behind both human cognitive functioning and for a design process that is associated with a high level of learning outcome. The paramount question then becomes; how can we design experiences that allow learners to experiment with knowledge in a setting or context that is controllable, encouraging them to form connections by experiencing a wide range of experimental possibilities around any given piece of information?

Today, gaming is seen by many as a media for pure entertainment purposes, with few or no learning benefits associated with it, and something which today's youths already spend too much time on. Incorporating learning objectives may then be seen as a way of making gaming more socially accepted and a further excuse for engaging in the activity. However, research has indicated that gaming, even without any explicitly expressed learning objectives, may have some potentially positive side effects. A recent study performed by Rosser et al. (2007) showed that surgeons who played video games for three hours a week had 37 % fewer errors and accomplished tasks 27 % faster, despite the fact that the observation was based on the trivial video game 'Super Monkey Ball'. If games produced for the sole purposes of entertainment have positive side effects associated with them when it comes to skills training, imagine what games tailored for skills training could do to this form of learning. This line of research also highlights a very important issue concerning gaming and simulator training, namely that there is not an easy relationship between simulator training and learning. Simulators can not in themselves teach – they have to incorporate knowledge and accommodate a specified group of users, and even then there are

---

2   Fidelity refers to how closely a simulation imitates reality (Alessi, 1988).

no certainties concerning what people will learn. Humans are complex; some learn best while reading, some learn best by doing, and some will have trouble learning no matter what learning paradigm is used. This shows that there are few guarantees when it comes to this form of learning, or any other form, but trying to understand the underlying mechanisms of such training is likely to lead to a higher probability of success.

This paper will explore a more detailed picture of what simulator training entails in connection with the differentiation of various levels of human cognitive functioning. The Skills, Rules and Knowledge taxonomy (SRK) developed by Rasmussen (1983) is used as a framework for mapping human cognitive functioning and is introduced first in this review. A general discussion on the role of fidelity and transfer in simulator training is presented, in connection with the distinction between 'hard' and 'soft' skills, but also through showing that different kinds of cognitive functioning require different kinds of simulator interfaces and layouts. Following the discussion, some key principles for effective learning design will be presented.

2 Cognitive models

The distinction between the common understanding of the concept 'hard' and 'soft' skills has been made by several researchers on learning and cognition (Sun, Merill, & Peterson, 2001). Anderson (1993), Keil (1989) and others, have proposed a distinction between procedural and declarative knowledge with the former meaning what is commonly referred to as 'soft' skills or 'know-how' and the latter referring to 'hard' skills or descriptive knowledge. Smolensky (1988) proposes a distinction between conceptual (publicly accessible) and subconscious (inaccessible) processing. Yet another distinction has been made by Dreyfus & Dreyfus (1986) between analytical and intuitive thinking. This paper will use the distinction between procedural and declarative skills.

Further, this paper argues for the use of the SRK taxonomy as a proficient tool for classifying different kinds of simulators. We do not need to know all the processes that underlie all human cognitive functioning in order to hypothesize about simulator training. In essence, we only need to know that the processes are there, and that they fulfil some sort of function. This means that we need a model[3] which has a level of complexity that corresponds to the level of complexity of the phenomena we would like to say something about. Having a skewed relationship between the tool used and the phenomena we would like to hypothesize about will only add to the complexity and not necessarily add to the result. It is the claim of this paper that the SRK taxonomy has the appropriate level of complexity that makes it the most proper tool for this line of argument.

---

3   Model: "A representation that mirrors, duplicates, imitates or in some way illustrates a pattern of relationships observed in data or in nature …. A model becomes a kind of mini theory, a characterization of a process and, as such, its value and usefulness derive from the predictions one can make from it and its role in guiding and developing theory and research» (Reber, 1995).

Allan (1993) argues that there has been a dichotomy between the providers of content and the providers of technology-mediated learning environments, such as simulators, for too long. The result of such a dichotomy may directly affect the quality of the product by losing focus on what should be the most important aspect of such training, namely its ability to foster skills. However, arguing for a bifurcate between the content and the provider (in this case, a simulator), it is not this paper's claim that the SRK taxonomy provides a sufficient level of complexity for any and all forms of content.

On the contrary, applying a structural model such as the SRK taxonomy to a complex skill, such as for instance negotiation, would be too simplistic. The SRK taxonomy is not a model explaining complex psychological processes; it provides a useful set of categories of human performance. This will make the SRK taxonomy fall short in explaining underling processes of human performance. For making a simulator an appropriate tool for learning how to negotiate we need a more complex model for understanding the underlying psychological processes of human cognition. Such a model would need to be able to explain many of these processes in order for us to be able to simulate them. For examples of such models see, for instance, Beer (1985), Broadbent (1977), Craik (1943), Michon (1985), Miller, Galanter & Pribram (1960), Rasmussen (1983; 1986).

3 The skills, rules, knowledge taxonomy

The Skills, Rules, Knowledge taxonomy developed by Rasmussen (1983) is a suitable tool for mapping the different levels of cognitive functioning that have to be taken into consideration when designing and developing a training simulator. This taxonomy is useful not only for mapping cognitive functions, but also in defining different categories of human functioning and for understanding the different aspects of simulator training. Table 7.1 shows an overview of the relation between levels of cognitive control in the SRK taxonomy and the way in which constraints in the environment are represented and processed.

The SRK taxonomy consists of three levels, each defining a level of cognitive control (i.e. a different category of human action). At the top, knowledge-based behaviour (KBB) of cognitive control is defined by analytical reasoning that is serial and based on a symbolic representation of the relevant constraints in the environment (Vicente, 1999). By representing the goal-relevant constraint in the environment as a mental model, KBB guides action. The middle row in Table 7.1 shows the rule-based behaviour (RBB) level of cognitive control. This level is characterized by an if–then mapping between a familiar perceptual cue and the appropriate response. No reason is required and instead there is a direct link between the cue and the responses. RBB then guides actions by representing the perceptual constraints in the environment in terms of perceptually grounded rules (Vicente, 1999). The skill-based behaviour (SBB) level of cognitive control involves real-time coupling to the environment through what Rasmussen (1983) refers to as a 'dynamic real world model', which is an implicit model of the environment (Vicente, 1999), much

Table 7.1 *Relation between levels of cognitive control in the SRK taxonomy and the way in which constraints in the environment are represented and processed (adapted from Vicente (1999).*

Behaviour	Representation of Problem Space	Process rules
Knowledge-Based	Mental model; explicit representation of relational structures; part–whole, means–end, causal, generic, episodic relation, etc.	Heuristics and rules for model creation and transformation; mapping between abstraction levels; heuristics for thought experiments.
Rule-Based	Implicit in terms of cue–action mapping; black-box action–responses models.	Situation-related rules for operation in the task environment, i.e., on its physical or symbolic objects.
Skill-Based	Internal, dynamic world model representing the behaviour of the environment and the body in real time.	Not relevant. An active simulation model is controlled by laws of nature, not by rules.

like a physician learns through experience to recognize breathing sounds that indicate pneumonia without having to represent those sounds explicitly internally. SBB provides a basis for direct coupling and parallel, continuous interaction with the world (Vicente, 1999). This three-level taxonomy can be represented graphically, as shown in Figure 7.1.

SBB consists of automated, highly integrated and smooth patterns of actions that do not need any conscious attention. A typical example is walking, which is an automated psychomotor activity driven by a continuous perception-action loop (Vicente, 1999). SBB consists of anticipated actions, meaning that it can initiate actions before

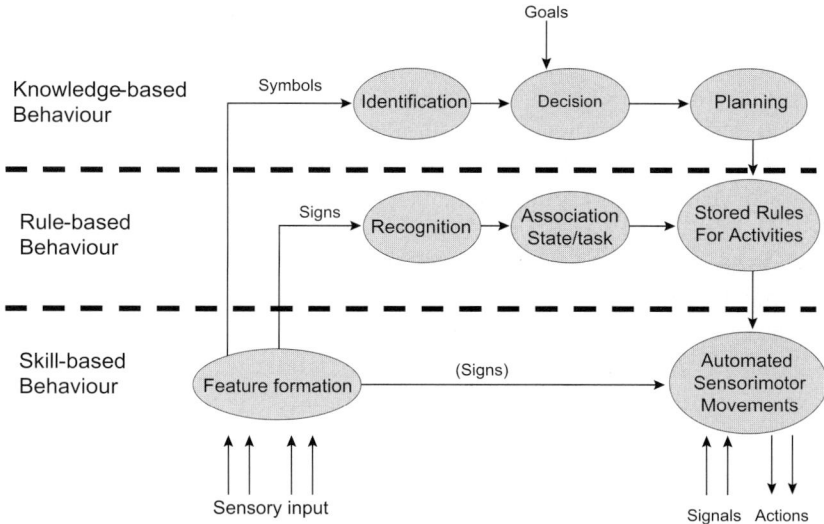

Figure 7.1 *The SRK taxonomy of human performance categories (adapted from Vicente 1999).*

the state of the world has actually changed. SBB cannot be verbalized because it does not require conscious attention (Vicente, 1999).

RBB consists of stored rules that derive from procedures, experiences, instructions, or previous problem-solving activities (Vicente, 1999). Action is goal-oriented, but goals are not directly represented. Workers may know the goal that the rules can achieve, but they do not reflect on those goals when following the rules. In RBB mode people do not reasoning, they merely follow familiar perceptual cues in the environment which trigger action directly. In contrast with SBB, people in RBB mode are usually aware of their cognitive activities and can therefore verbalize their thoughts (Vicente, 1999).

Last, KBB consist of deliberate serial search based on an explicit representation of the goal and a mental model of the functional properties of the environment. In contrast to RBB, the goals in KBB are considered explicitly rather than implicitly (Vicente, 1999). KBB is slow, serial and effortful because this mode requires conscious focal attention. KBB is often used in unfamiliar settings where previous experiences are no longer valid and solutions must be improvised by reasoning (Vicente, 1999).

4 The classification of simulators

To date, there is no paramount system for the classification of simulators, even though some have been proposed (Sulistio, Yeo, & Buyya, 2004). Simulators are often treated in discussions as a unitary concept and as a tool for facilitating learning, without a thorough understanding of the underlying structures that actually facilitate learning. Based on the distinction mentioned earlier it is possible to divide simulator training according to what form of skills such training is mainly meant to foster, namely simulators relying on declarative knowledge and those relying on procedural knowledge. This distinction represents a major difference, not only in terms of human cognitive processing, but also it should represent a huge difference when it comes to designing and developing simulators.

As representative of the former kind of simulators, this paper will use flight simulators as a basis for discussion throughout. This is in order to exemplify the differences between simulators appropriate for fostering rule-based behaviour on the one hand and knowledge- and skill-based behaviour on the other hand. This does not mean that this paper does not acknowledge that all three kinds of skills are present during flight, rather the intention is to illustrate the difference of the rule-based part of flight simulator training, such as e.g. the actual flight handling, and not the whole process from take-off to landing that would include procedures such as Air Traffic Control correspondence and weather considerations.

Aldrich (2005) divides computer-based simulators into four main categories, based in part on their appropriateness for training in different kinds of business skills.[4] In *Branching Stories* the user makes multiple-choice decisions along an

---

4   All simulators referred to in this article are computer-based.

ongoing sequence of events as to what to say to another person in a given situation. This interaction influences the decisions made under way and ultimately the outcome of the interaction. *Interactive Spreadsheets* focus on abstract business school issues such as supply chain management, product life cycles, and accounting. The task may be, either alone or as teams, to allocate some sort of resources along a turned based and fixed interval, and then see the result play out on dense graphs and charts. In *Game-based Models* the user engages in familiar games such as solitaire or memory, but with important pieces of linear or task-based content replacing trivia or icons. In *Virtual Labs/Virtual Products* the players interact with visually accurate representations of actual products without the physical restrictions of reality. The interface aligns with the real functions of the object represented. The *Virtual Products* forsake some of the fidelity of *Virtual Labs* by focusing instead on the situation in which the product is being used.

Alessi (1988) makes another distinction between situational, procedural, process, and physical simulators. *Procedural* and *Situational* simulators refer to simulators which have as their primary objective to teach someone *how to do something*. For instance, an example of a *procedural* simulator is a flight simulator, while an example of a *situational* simulator may be simulations for classroom management. *Physical* and *Process* simulators are simulators that have as their primary objective to *learn about* something. An example of a *physical* simulator may be a simulator that simulates a phenomenon or a physical object and its behaviour. In *Process* simulators, on the other hand, the phenomena to be simulated are invisible processes, such as genetics or population growth.

This paper will focus on yet another distinction, namely the levels of cognitive functioning involved in mastering the simulator. Through identifying the level of cognitive involvement within the user, it is possible to develop and design simulators that are tailored for different kinds of users, in different kinds of environments, and with different kinds of objectives.

5 The role of fidelity and transfer in simulations

A common misconception when it comes to fidelity is the belief that the higher the fidelity there is between the learning situation and the application situation, the higher the transfer.[5] This notion implies that some sort of linear relationship exists between fidelity and transfer. However, research indicates a more complex non-linear relationship (Alessi, 1988). Even if a straightforward relationship does not exist between fidelity and transfer and that this relationship is difficult to describe, we cannot simply deny that such a relationship exists and from that conclude that this relationship is not important. On the contrary, fidelity is important, but sometimes less can be more in the sense that a higher degree of fidelity does not automati-

---

5   Transfer of learning refers to an individual's ability to apply something learned in one situation and transfer that knowledge to another novel situation (Singley & Anderson, 1989).

cally ensure a higher degree of transfer. Previous research in the field has revealed mixed results when comparing fidelity and transfer. Some studies have even shown there are no differences in learning or transfer due to fidelity (Cox, Wood, Boren, & Thorne, 1965, cited in Alessi, 1988). Other studies have shown that simulators can teach more effectively than real airplane flight, and that students training in flight simulators require fewer hours to reach the requirements needed than those who do not receive such training (Alessi, 1988; Povenmire & Roscoe, 1973; Roscoe, 1971; Valverde, 1973). Due to the fact that real flights have the highest fidelity, this must mean that lower fidelity sometimes is more effective.

The relationship between fidelity and learning is complex but there are some tendencies. For very high fidelity the amount of learning can decrease, and this is especially the case with less experienced students. Putting a novice in a high fidelity simulator or even in real flight (i.e. the highest possible level of fidelity) would be too stressful and even confusing, producing no learning at all. At the other end of the continuum, putting an experienced student in the same situation would produce more learning than presenting him or her with an instructional video or a written text on piloting.

So, how does fidelity relate to transfer? One of the main problems related to the transfer of skills lies in the fundamental educational question, namely that it is rare that people learn things in school which apply directly to life and work. Transfer fails when knowledge required in one situation fails to transfer to another, and a major theoretical question arises as to why this is so. One plausible explanation is that it is the result of the inevitable consequence of the limited power and generality of human knowledge (Singley & Anderson, 1989). Just to have the knowledge that logically implies a solution to a task is not enough. One must learn how to apply that specific knowledge to a specific task in a specific situation. Furthermore, the probability of transfer decreases as the time interval between the original task and the transfer task increases (Singley & Anderson, 1989). This point is also likely to be connected to retrieval by the notion that information that has been acquired recently is more likely to be retrieved than information acquired further back in time (Ormrod, 2004). The concept of transfer is a complex one, containing several aspects such as the similarities between the performance environment and the instructional environment, and the perceived similarities between the performance environment and the instructional situation and the learner's level of overall motivation. These factors are not independent. Actual similarity affects perceived similarity, which in turn affects motivation. In sum, it seems we are at an impasse in simulation design (Alessi, 1988). Increasing fidelity, which theoretically should produce a higher degree of transfer, may on the contrary inhibit learning, which in turn will inhibit transfer. On the other hand, decreasing fidelity would increase learning, but what is learned may not transfer to the application situation if those settings are too dissimilar.

6 Implications for simulator training

One of the main objectives for a flight simulator is to put the user in a life-like environment in terms of visual and kinetic aspects of flight, where the pilots can train on different scenarios associated with piloting an aircraft (Farmer, Rooij, Riemersma, Jorna, & Moraal, 1999; Salas, Bowers, & Rhodenizer, 1998). This is not to say that a pilot does not have to rely on knowledge-based skills. When the pilot encounters unforeseen events, he/she has to rely on his/her KBB skills a great extent in order to come up with a good solution for any given problem. This kind of simulator lies in the interchange between RBB and KBB.

When encountering a given problem, the pilot's actions and responses often follow a strict and rigid pattern which is often sequential and deterministic in the sequence in which they have to be executed. When the pilot is confronted with the problem 'fire in the left engine', he or she has to engage in a sequential pattern of responses in order to avoid a disaster. This pattern of responses is not open for much individual interpretation by the pilot in terms on what the best course of action is. The obvious reason for this is that it is both time-efficient and to a greater extent ensures a successful outcome to the problem. For safety reasons, pilots do not have the luxury of considering multiple options to a problem, but have to act according to the sequence that has been tested and found to have the highest success rate. Different sequences to different problems are carefully described in manuals that the pilot must be familiar with.

Simulators for implicit or procedural skills are associated with behaviours such as KBB and SBB in the SRK taxonomy, examples of which are negotiation, leadership and decision making. The traditions of simulating these types of skills have been less prevalent and have not been given the same attention up until now (Aldrich, 2004; 2005; Crookall & Arai, 1995; Quinn, 2005). There are no clear-cut answers as to why this is so but some possible explanations will be presented here. To simulate negotiation, for instance, there can be no fixed pattern of responses to ensure a favourable outcome. In situations where the interaction with another person is the task, expecting uniform responses is intrinsically unrealistic. An almost infinite number of responses posted at almost an infinite number of places in the process could theoretically lead to a favourable outcome (Suchman, 1987). What is classified as a favourable outcome could also fluctuate greatly between players as they interpret different situations differently. KBB in the form of cognitive control is defined by analytical reasoning that is serial and based on a symbolic representation of the relevant constraint in the environment. Negotiation is a highly dynamic process which focuses on different aspects of cognitive processing. Unlike negotiation, SBB consists of automated, highly integrated and smooth patterns of actions that do not need any conscious attention,| such as walking, and they are often recognized by the lack of direct consciousness. This kind of psychomotor activity is driven by a continuous perception-action loop (Vicente, 1999).

In simulators focusing on RBB and SBB the learning benefits are to a large extent associated with the repetition of fixed responses to a number of problems that to a great extent can be revealed beforehand. Pilots are supposed to have 'spinal reflexes' to a variety of problems that can emerge during a flight. In addition, simulating mechanical malfunctions is more transparent because machines are bound to certain rules. The range of possible malfunctions is limited and easier to simulate than human behaviour that is not bound by the same narrow set of rules and therefore much more complex. In a simulator for KBB and SBB, such as one for negotiation, the number of possible responses exceeds the number that reasonably can be implemented in a simulator. Therefore the benefits of such a simulator depend to a large extent on the reflections on which course of action to follow and on what outcomes follow which courses of action, more than on the repetitions of sequences.

One of the hypotheses on the learning benefits associated with RBB simulator training lies in the notion that life-like 'micro world' simulators make the transition from a training environment to the real world easier and more trouble free (Singley & Anderson, 1989). The reasons for that is the belief that the more familiar the pilot is with the environment he or she is supposed to operate in, such as the cockpit, the faster he or she can adapt to a variety of different situations. The rationale behind this assumed relationship between familiarity and reality might in part be explained by a 'situated' framework for learning, where some cognitivists have proposed that most learning is context specific and that it is 'situated' in the environment in which it takes place. Such situated learning is unlikely to result in transfer to new contexts, especially when they are very different from the ones in which learning originally occurred (Lave & Wenger, 1991; Ormrod, 2004; Singley & Anderson, 1989; Suchman, 1987). As long as two tasks have something in common, the possibility of transfer from one situation to another exists (Ormrod, 2004). However, commonalities among tasks do not guarantee transfer and, regardless of theoretical orientation, there is a wide agreement among scientists that transfer does not occur as often as it should or could (Ormrod, 2004).

This area of research indicates the importance of treating simulations as a dynamic concept and that each simulator needs a high degree of customization. This can be done in part by addressing and analysing the individual user's level of competence, and varying the instructional level. Developing a simulator under the slogan 'one size fits all' could fail to meet many desired criteria needed to obtain any preferred level of learning outcome.

When training on in-flight simulators pilots are bound to follow strict protocol and, for the most part, the skills that must be mastered can be expressed verbally or through guidance from a more competent person. This means that there is a fairly lucid relationship between the skills that have to be mastered and the degree of externalization needed in order to master that particular skill, and the learning outcome is associated with repetition of these skills. However, in simulators were this relationship is not as straightforward, such as a simulator for negotiation, one of the challenges is to be able to make internalized skills externalized. This argument assumes that everyone, through human interaction, has some basic skills in negotiation, and that some are better and more experienced than others. The learning outcomes associated with negotiation training in a simulator are not closely tied to actual motor performance, but lie in the interchange between being self-aware and being able to meta-reflect. There is no *one* correct way of doing things, and therefore the correctness may lie in the *awareness* of why one chooses to act in the way one does as much as the act itself. However, in a flight simulator it is possible to pinpoint right from wrong lines of action to a greater extent. This constitutes a marked difference between simulators designed for declarative and motor skill training and simulators for procedural skill training, and should be accounted for in the design and when deciding on the level of fidelity.

7 Design and implications for learning

When designing learning environments, such as simulators, there is a magnitude of design decisions to be made. During such a process many of the design issues are made unconsciously along the way, without a thorough understanding of what the decisions can lead to in terms of the trade-offs involved. One possible reason for this is that there may have been a dichotomy between the designers of simulator environments (often engineers) and those providing content (often educationalists) in the sense that designers and those in charge of the content may not necessarily have had concurrent goals or focus when designing such environments (Allan, 1993).

In designing technology-mediated environments the first thing designers should ask themselves is what the goals are: What explicitly is to be learned? The answers to such questions should, to a great extent, influence the design process, and there should at all times be a close link between the providers of design and the providers of content. More explicitly, in a design process the following, based on the work of Allan (1993), should be taken under consideration.

Memorization versus thoughtfulness: Should the focus be on memorizing certain traits/facts, or is the main goal to stimulate the ability to solve complex problems through the ability to reflect, communicate, and engage in cooperative behaviour?

Whole tasks versus component skills tasks: There is a trade-off between having the focus on whole tasks and component skill tasks. Is the environment meant to stimulate the learning of whole tasks that require the integration of a variety of skills, or is the environment meant to stimulate the acquisition of simplified tasks and focus on particular subskills?

Breadth or depth of knowledge: Is the primary focus of the simulator to stimulate the acquisition of a little knowledge of a lot of things, or is the focus the deep and thorough understanding of a few topics?

Cognitive versus physical fidelity: When designing and creating simulated environments a critical question emerges in the trade-off between preserving physical fidelity to the environment and preserving only the cognitive fidelity. This difference can be illustrated by the difference of having a simulated control-room in scale 1:1, identically designed with buttons and flow charges which have to be physically operated by people, or have the same configurations of the entire system and subprocesses represented on a computer screen.

Incidental versus direct learning: When putting someone in a learning environment, can what you want them to learn be taught directly, through the specific tasks they engage in, or can it be taught incidentally to the task?

Learner control versus computer control: There is a trade-off between putting the learner in control of his or her learning versus letting the computer control the whole process of learning.

Another part of simulator training that has not been given much attention is that in designing simulators there has been little focus on designing environments fostering the collaboration of several people who are interacting simultaneously (Dalziel, 2003). One tradition within designing learning environments that has attempted to put focus on these topics is Learning Design.

Learning Design has emerged as one of the most significant recent developments within the broader e-learning paradigm (Dalziel, 2003). Learning Design's core principle places focus on the relationship between learner and content and learner and peer or instructor. There are several different definitions of Learning Design, but common to all is the greater focus on content and activity of the learner, and greater emphasis on the multi-learner as opposed to the single learner (Dalziel, 2003).

One of the problems concerning design is that too often content has been considered to be the primary focus, thereby defining the course or the way content is presented (Sims, 2005). The goal of Learning Design and simulator training should be to arrange for activities designed to engage with key elements of the course content. In this way the learner becomes integrated with a situated and contextualized environment, providing them with a level of control over that learning environment (Sims, 2005).

8 Summary

In search of a more even debate of the use of simulators for skills fostering, this review paper has put forward a possible way of viewing simulator training by arguing the need to start with classifying the skills we wish to foster in terms of cognitive functioning before hypothesizing about what kinds of simulators that are best suited for the actual skills fostering. Applying the SRK taxonomy as a framework for such a classification is useful. The SRK taxonomy has a proficient and suitable level of complexity to function as a framework for such a classification in that it provides a functional dismemberment of human cognitive functioning. This possible classification has been seen in close relationship with both fidelity and transfer in order to be able to draw some conclusions as to what kind of simulators are best suited to foster different kinds of skills. Finally, this paper has presented some general principles related to the design of such learning media, in relation to the principles discussed earlier in this paper.

9   Concluding remarks

The acquisition and use of skills constitutes a major portion of human activity. As we develop as human beings our methods for skills acquisition evolve with us. All evolution does not necessarily constitute improvement. Simulation training as a method for skills acquisition is an evolutionary improvement of modern times. However, embracing simulation without having a thorough understanding of the mechanisms involved can result in a false belief about the efficiency of this method. It is important to acknowledge that it is necessary to abandon the classical view of learning as the accumulation of facts and accept that learning is not about explicit knowledge but about getting people to understand what is going on, to reflect on different courses of action, and to be able to act based on these reflections. Traditional teaching argues that we can only teach what we know, meaning we can only teach what is explicitly accessible from our consciousness. This means that a great deal of what we know but not are able to express, cannot be taught in terms of traditional methods. This opens for a view of learning as something one has to engage in, using tools such as simulators.

People have some profound misconceptions about what it means to know (Schank, 2002). Such misconceptions come about because facts are what people believe it is to 'know'. That is a grave misconception because we do not know what we know because so much of what we know is tied up in various schemas and in cognitive strategies making sure that our perceptual and cognitive systems do not overload. This enables us to function in a complex world rather effortlessly.

People do not know how they learn, how they understand, or how they came about believing what they believe. We can at all times do things, perform and behave in our daily life without being able to explicitly state the rules that govern this behaviour, the knowledge concerning these domains is simply not in our consciousness. The question then arises as to how we can exploit this 'unknown', implicit knowledge.

An important aspect of classical views of teaching is that one only teaches what is testable and the question easily becomes a question of what we can test rather than what is the best teaching. Given the premises that we can only teach what we know, and that what we know in terms of explicit knowledge is just a very small part of what we can call overall knowledge, and also the fact that classical teaching is viewed by many as being archaic, training in a simulator can surpass this problem by providing learners with a environment that fosters the creation of their own learning experience. By the active manipulation of such an environment and by learners becoming their own teachers within a framework of 'learning by doing', I strongly believe that a negotiation simulator has the potential to be a useful tool for learning.

References

Aldrich, C. (2004). Simulations and the future of Learning. San Francisco: John Wiley & Sons, Inc.
Aldrich, C. (2005). Learning by Doing. San Francisco: Pfeiffer.
Alessi, S. M. (1988). Fidelity in the design of instructional simulations. Journal of Computer-Based Instruction, 15(2), 40-47.
Allan, C. (1993). Design Issues for Learning Environments (No. Technical Report No. 27). New York: Centre for Technology in Education.
Anderson, J. (1993). Rules of the mind. Hillsdale: Lawrence Erlbaum Associates.
Beer, S. (1985). Diagnosing the system for organizations. Chichester, UK: Wiley.
Broadbent, D. E. (1977). Levels, hierarchies, and the locus of control. Quarterly Journal of Experimental Psychology, 29, 181-209.
Cox, J. A., Wood, R. O., Boren, L. M., & Thorne, H. W. (1965). Functional and appearance fidelity of training devices for fixed-procedural tasks.
Craik, K. J. W. (1943). The Nature of Explanation. Cambridge, UK: Cambridge University Press
Crookall, D., & Arai, K. (1995). Simulation and Gaming across Disciplines and Culture: ISAGA at a Watershed. London: Sage Publications Ltd.
Dalziel, J. (2003). Implementing Learning Design: The Learning Activity Management System (LAMS). Paper presented at the Australasian Society for Computers in Learning in Tertiary Education (ASCILITE), Adelaide, Australia
Dreyfus, H. L., & Dreyfus, S. E. (1986). Mind over Machine. The Power of Human Intuition and Expertise in the Era of the Computer. New York: The Free Press.
Farmer, E., Rooij, J., Riemersma, J., Jorna, P., & Moraal, J. (1999). Handbook of Simulator-Based Training. Burlington, USA: Ashgate Publishing Limited.
Keil, F. (1989). Concepts, kinds and cognitive development. Cambridge: MIT Press.
Lave, J., & Wenger, E. (1991). Situated Learning: Legitimate peripheral participation. Cambridge: Cambridgen, University Press.
Michon, J. A. (1985). A critical view of driver behaviour models. What do we know, what should we do? In L. Evans & R. Schwing (Eds.), Human behaviour and traffic safety (pp. 485-525). New York, NY: Plenum Press
Miller, G. A., Galanter, E., & Pribram, K. H. (1960). Plans and the structure of behavior. New York, NY: Holt, Rinehart & Winston
Ormrod, J. E. (2004). Human Learning. New Jersey: Pearson Prentice Hall.
Povenmire, H., & Roscoe, S. (1973). Incremental transfer effectiveness of a ground-based general aviator trainer. Human Factors, 15, 534-542.
Quinn, C. N. (2005). Designing e-Learning Simulation Games. San Francisco: Pfeiffer.
Rasmussen, J. (1983). Skills, rules, and knowledge; signals, signs, and symbols, and other distinctions in human performance models. IEEE Transactions on systems, man and cybernetics, 13, 257-266.
Rasmussen, J. (1986). Information processing and human-machine interaction: An approach to cognitive engineering: New York: Elsevier science.
Reber, A. S. (1995). The Penguin Dictionary of Psychology: Puffin Books.
Roscoe, S. (1971). Incremental transfer effectiveness. Human Factors, 13, 561-567.
Rosser, J., Lynch, P., Cuddihy, L., Gentile, D., Klonsky, J. J., & Merrell, R. (2007). The Impact of Video Games on Training Surgeons and Physicians in the 21st Century. Archives of Surgery, 142(2), 181-186.
Ruben, B. D. (1999). Simulations, Games, and Experience-Based Learning: The Quest for a New Paradigm for Teaching and Learning. Simulation & Gaming 30(4), 498-505.
Salas, E., Bowers, C. A., & Rhodenizer, L. (1998). It is not how much you have but how you use it: Toward a rational use of simulation to support aviation training. International Journal of Aviation Psychology, 8(3), 197-208.
Schank, R. C. (2002). Designing World-Class E-Learning. New York: McGraw-Hill.

Sims, R. (2005). Beyond Instructional Design: Making Learning Design a Reality. Journal of Learning Design, 1 (2), 1-9.

Singley, M. K., & Anderson, J. R. (1989). The Transfer of Cognitive Skill. London: Harvard University Press.

Smolensky, P. (1988). On the proper treatment of connectionism. Behavioral and Brain Sciences, 11(1), 1-74.

Suchman, L. (1987). Plans and Situated Actions. Cambridge: Cambridge University Press.

Sulistio, A., Yeo, C., & Buyya, R. (2004). A taxonomy of computer-based simulations and its mapping to parallel and distributed systems simulation tools. Software-Practice and Experience, 34, 653-673.

Sun, R., Merill, E., & Peterson, T. (2001). From implicit skills to explicit knowledge: a bottom-up model of skill learning. Cognitive Science, 25, 203-244.

Valverde, H. (1973). A review of flight simulator transfer of training studies. Human Factors, 15, 510-523.

Vicente, K. J. (1999). Cognitive Work Analysis : Toward Safe, Productive, and Healthy Computer-Based Work: LEA, Inc.

8

User-Centred Development of Simulator-based Training – An Exploratory Case Study

Kjell-Morten Bratsberg Thorsen

The development of training simulators poses new requirements for User-Centred Design methods, as the simulator will support a different context of use than the one analysed to inform its design. These requirements were explored in a case study at the University of Oslo, by applying existing user-centred methods to the design of a simulator for negotiation training, leading to a suggestion for a possible approach to the design of simulator-based training, founded mainly on Contextual Design and Cognitive Work Analysis, and a formative approach to User-Centred Design.

1 Introduction

Many attempts have been made at using technology to facilitate learning – with various degrees of success (Koschmann, 1996). Perhaps one of the more successful and promising attempts is the use of simulators to provide training in fields where real training is too expensive, dangerous, or simply not practical, of which military settings (Farmer, Rooij, Riemersma, Jorna, & Moraal, 1999) and aviation (Salas, Bowers, & Rhodenizer, 1998) are typical examples. Traditionally, the skills learned through simulation have mostly been based on physical or procedural operations, but now we see the advent of simulators for training in more social or 'soft' skills, such as leadership (Aldrich, 2004). This poses a special challenge in the development of simulators, as human behaviour and social interaction have to be part of the simulations, in addition to the rules of physics.

A natural choice for the development of any training simulator, and especially one for training in social skills, would be the User-Centred Design (UCD) approach, with its main tenet to keep the users, not the technology, at the centre of development (Faulkner, 2000; ISO 13407, 1999; Nielsen, 1993). In principle, UCD methods may be applied to the development of any system with human users. However, in most cases, UCD approaches have the development of tools to support work as their focus (Beyer & Holtzblatt, 1998; Vicente, 1999; Woods & Hollnagel, 2006). There is an underlying premise that a system designed on the basis of data of how users perform relevant tasks in a specific context will be fed back to that context of use to support and transform these tasks (Carroll & Campbell, 1989), and the development of training simulators does not fulfil this premise. The purpose of a training simulator is not to support work, but to simulate it as an environment for training and exploration (Gredler, 2004). Hence, the development of a training simulator relates to two distinct contexts of use – the one that will be simulated and offered in training, and the one in which the actual training will occur. The design of a simulator will be based on data from the former context with the purpose of supporting the activity in the latter. In other words, the analysed context and the context in which the simulator will be used are not the same, and the development process has to relate to both.

In order to explore this gap between traditional UCD methods and the situation of designing a training simulator, an exploratory study was conducted within a project aimed at developing a simulator for negotiation training. The development project served as a testbed where user-centred methods and techniques were applied, and the activities of the design team were recorded, reflected on, and later analysed to generate ideas for a User-Centred Design method for training simulators. The *Human-Centred design processes for interactive systems* (ISO 13407, 1999) provided an overarching framework for the design process, within which conventional user-centred techniques from the Usability Engineering approach and the more specialised method Contextual Design (Beyer & Holtzblatt, 1998) were applied.

This paper presents the methodological background from User-Centred Design and training simulator development, and describes and discusses the relevant activities and results from the exploratory case study. Finally, based on tendencies and requirements discovered during the study, a possible approach to a design method for the development of simulator-based training is suggested. The emphasis will be on design methods and techniques, and pedagogical issues will only be mentioned briefly.

2 Background

2.1 User-Centred Design

The term 'User-Centred Design' refers to a general approach to artefact and system development, where two of the most central principles are to involve users directly or indirectly in the design process, and to design, test and redesign the system or artefact incrementally through an iterative process (Faulkner, 2000; Gould & Lewis, 1985; ISO 13407, 1999; Karat, 1996; Mayhew, 1999; Nielsen, 1993). The approach has its origins within Human Factors, also known as Ergonomics, a multidisciplinary field spanning disciplines such as psychology, informatics and sociology, to name some, and has the design of artefacts compatible with human properties as its main focus (Vicente, 2004).

The International Organization for Standardization (ISO) has formulated the *Human-centred design processes for interactive systems* (ISO 13407, 1999) as a guiding framework for User-Centred Design, and a similar process is presented in the Usability literature as the *Usability Engineering Lifecycle* (Faulkner, 2000; Mayhew, 1999; Nielsen, 1993). ISO 13407 (1999) is only intended to complement existing methods and techniques, while Usability Engineering as a field includes a multitude of possible techniques for all phases (Faulkner, 2000; Kuniavsky, 2003; Mayhew, 1999; Nielsen, 1993), based on the definition of Usability as the extent to which a system is easy to learn, easy to remember, efficient and pleasant to use, and whether it has a low error rate (Nielsen, 1993).

Both processes start by analysing the users, the tasks they perform, and the context in which they occur, summarised by ISO as a phase for understanding and specifying the context of use. This analysis leads to the specification of requirements and goals for the system. In the next phase, design solutions are produced by iterating through a design–test cycle where prototypes or mock-ups are continuously evaluated and tested with actual or potential users. Finally, the designs are evaluated against the requirements to assess whether these have been fulfilled. If they have, the design is ready for production, but if not, then another iteration through the cycle is needed. The iterative nature of the process implies that not only the design but also the requirements may be changed and refined throughout the development process. This is in contrast to the more traditional development processes, where the requirements are formulated once, at the beginning of the process, and only the finalised product is tested and evaluated against them (Sommerville, 2004).

While the User-Centred Design process may offer solutions to many of the challenges of traditional development processes, challenges still remain. One such challenge lies in the inherent assumption that the developed artefact is intended to support tasks in the context of use that was initially analysed, frequently referred to as the *task–artefact cycle* (Carroll & Campbell, 1989). The introduction of a new artefact inevitably transforms the original task, as it both enables and compels new ways of working, and in effect, the artefact will be used to support a different task than it was initially designed for. Further iterations may refine the artefact accordingly, but the redesigned artefact will again transform the task, and so the artefact will always be one step behind. Vicente (1999) has ascribed this problem to a *descriptive* approach to work analysis, where the way a system *does* behave is studied. Instead, Vicente proposes a *formative* approach for analysing the way a system *could* behave, by focusing on the boundaries or constraints for work, independent of current work practice and tools. The descriptive techniques may still be used as a basis, but data from these analyses are further generalised into models of the underlying constraints of work, rather than descriptions of the task at the time of analysis. While the techniques of Usability Engineering are examples of the descriptive approach, examples of the formative approach may be found within *Cognitive Engineering* (Norman, 1986). The two fields are highly related and partly overlapping, but while the former is mostly concerned with the design of artefacts that are usable for a single human user, the latter focuses its analysis and design on sociotechnical systems comprising both humans and artefacts, as well as their interaction (Hollnagel, 1998; Vicente, 1998; 1999; Woods & Hollnagel, 2006).

Cognitive Work Analysis (Vicente, 1999) is an example of a formative approach concerned with the analysis of complex sociotechnical systems. In the tradition of Cognitive Engineering, the focus of Cognitive Work Analysis (CWA) is not on the design of an artefact only, but a new sociotechnical system that will induce and support future work practice. The analysis is based on five conceptual distinctions with direct implications for design: the Work Domain, Control Tasks, Strategies, Social-Organisational, and Worker Competencies. Instead of describing the current way of working, modelling tools are provided to find the intrinsic work constraints for each distinction, both technological and organisational. The resulting models, starting with the Work Domain, put further constraints on the degrees of freedom for action in the studied system. This leads to a definition of the boundaries for an end-user's actions in the system, without specifying the details of these actions. Furthermore, the analysis builds on an ecological, as opposed to a cognitive, approach to Human Factors. Rather than starting with the constraints following from user characteristics, as is often done in other user-centred methods, the ecological approach starts with the constraints imposed on the users by the environment. Instead of supporting a user's potentially faulty model, the system is designed to give the users a more realistic mental model (for an introduction to mental models in design see Norman, 2002).

Another formative method is Contextual Design (Beyer & Holtzblatt, 1998; Vicente, 1999), consisting of multiple user-centred techniques collected as an integrated process and tailored to the challenges of working in design teams. Actually, Contextual Design (CD) uses the term 'customer' rather than 'user', to make a point of the fact that not only direct users of a system are affected by it, but other than this the terms are interchangeable. CD comprises five work models as a basis for the analysis and design process: a Flow Model, a Sequence Model, an Artefact Model, a Cultural Model, and a Physical Model. What makes this a formative approach is mainly the consolidation of sequential, story-based accounts of users' work practices into the structural work models and a hierarchical organisation of key statements from the accounts into an Affinity Diagram. Similar to CWA, the method is based on the principle of designing a new way of working based on technology. Several successive techniques lead to the design of a new work practice as a response to the user data, including drawing a sketch of the design as an overall picture of the new system, in CD referred to as a 'vision'. From there, the system is designed through an iterative process with rapid evaluation of prototypes with users, much like common user-centred prototyping techniques.

2.2 The Development of a Training Simulator

The term 'simulator' in this paper is to be understood as a system that is able to present a controlled representation of the appearance and/or behaviour of certain aspects of a real system and possibly also its environment (Farmer et al., 1999). Furthermore, the term 'system' is used in a broad sense, potentially including physical, technical, and social aspects, compliant with the term's use within General Systems Theory (Bertalanffy, 1972). However, a training simulator, as described in this paper, will most likely be implemented as a computer system.

In the case of a training simulator, the simulated aspects are chosen on the basis of intended learning outcomes and the user will be able to manipulate the simulation in a way that promotes practice and learning (Gredler, 2004). This is based on a constructionist view of learning, where one learns through an active and social process of constructing meaning (Kafai & Resnick, 1996; Koschmann, 1996; Squires, 1999). The simulator will present a context that the learner may actively engage in, to build understanding of the real context. Also, there should be a certain degree of correspondence between the simulation and the real system, an attribute referred to as *fidelity*, in order to support *transfer* of the acquired skills – meaning that the skills may be applied to a real-life situation (Alessi, 1988; Farmer et al., 1999).

As already mentioned, the development of a training simulator will involve two systems or contexts of use. There is the field of practice to offer training in, which will be referred to as the system to be simulated, and the system in which the simulator-supported activity of training will occur, referred to as the system to be supported. In the tradition of Activity Theory (Kuutti, 1996), the two systems may be represented as two different, but related activities, according to the basic structure of an activity depicted in Figure 8.1. In the system to be simulated, which in the study presented in this paper is the practice of negotiation, a negotiator – the *subject* of the activity

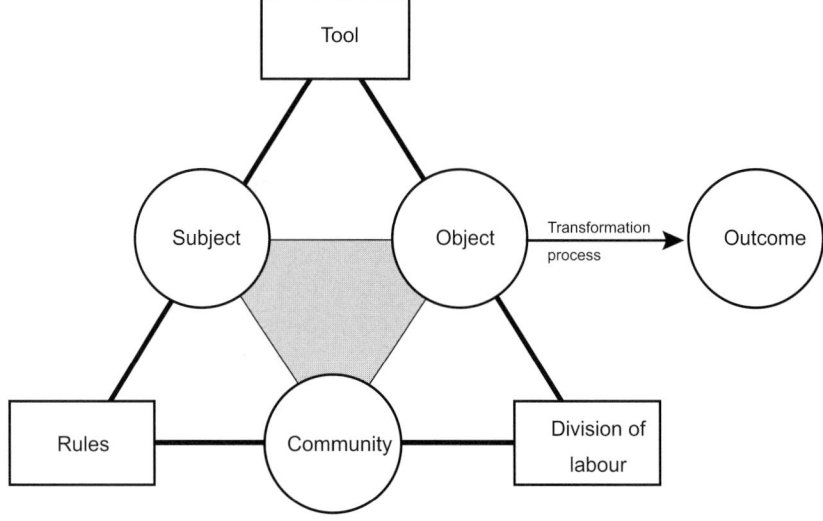

Figure 8.1 Basic structure of an activity (reproduced from Kuutti 1996).

– is using language, rhetoric, different strategies, etc., as *tools* to reach an agreement with the opponent, this agreement being the *object* of the activity. A negotiation happens within a *community*, possibly involving several team members from each party as well as other influenced persons and parties. Their interaction is controlled by *rules*, both formal and informal, and they have a *division of labour* in relation to the agreement to be achieved. In a similar vein, the training activity consists of a learner – the subject – within a community of other learners, instructors, and moderators, where the simulator is the tool that mediates the subject's training in negotiation, with the expansion of his or her negotiation skills as the object. Again, there are rules for the training, and with their different roles in the activity, learners, instructors, and moderators may be said to have a division of labour between them. While the details of these two activities are not important to this study, their existence is.

There are several relations between these systems or activities. First, the activity of negotiation – the system to be simulated, is represented in the training activity through the simulator, as part of the tool. Second, the intended outcome of the training is that the subject's negotiation skills may be transferred to the activity of negotiation. Third, the two activities are overlapping, as the learner will be involved in the activity of negotiation within the training activity, although in a simulated way. As an illustration of the latter point, a learner will simultaneously act with the intention of reaching an agreement with the simulated opponent and the intention of learning to negotiate, he or she will have to relate to both the rules of negotiation and the rules of the training situation, and so forth.

From this, it should be clear that the underlying premise of User-Centred Design that the artefact is intended to support the analysed activity does not hold in the case of developing a training simulator. Although the User-Centred Design processes presented in this paper are open frameworks and different techniques may be applied within them depending on the details and demands of each development situation, they do not give any guidance on how to include both the system to be simulated and the system to be supported in the process. Given these systems' different relations to the simulator to be designed, special clarification of their roles in the development is needed.

While no user-centred approaches to the development of training simulators has been formulated, a methodology for specifying the requirements for training simulators has been developed as part of a large European military programme, the MASTER project (*Military Application of Simulation and Training concepts based on Empirical Research*), to enhance the effectiveness and efficiency of military training (Farmer et al., 1999). The research group of the MASTER project observed that the requirements used to develop systems were also used to simulate them for training purposes, and instead the group proposed a methodology based on the principle that the requirements should be derived from training considerations. The methodology starts with a *Training Needs Analysis*, where the activity of the real system, the tasks performed by actors within this activity, the skills required for these tasks, and the trainees' existing skills are analysed. The discrepancies between the required and available skills are formulated as a set of training objectives, and provide a starting point for the next phase, the *Training Programme Design*, where the overall training programme, the specific training activities, training scenarios, and instructions are worked out. Finally, the detailed requirements for the training simulator are specified in the *Training Media Specification* phase. The MASTER approach is not built on the principles of User-Centred Design, but it is the only approach to acknowledge the two systems involved in training simulator development. In the first phase, the system to be simulated is analysed, and in the second phase, the system to be supported is designed. The third phase is mainly concerned with the simulator itself.

Furthermore, an account of the development of a simulator for leadership training has been given by Aldrich (2004). This is not a description of a development approach, user-centred or otherwise, but a case study of the trial and error process of designing and deploying this simulator. Aldrich and his colleagues started with a thorough analysis of leadership through literature reviews and interviews with leaders, resulting in their own model and theory of leadership. The aim of the analysis was to discover content that could be both cyclic and open-ended, meaning that it could be practised over and over and without a predefined outcome. They then decided on an interface for the simulator, the interaction it would provide, and the simulated settings in which to practice leadership. With this in place, the next step was to specify the rules for the inner workings of the simulator; the dialogue system, the artificial intelligence of the virtual characters, and the physical simulations and animations. Finally, the simulator was programmed and put to use. An important principle throughout this process was that every part of the simulator would reflect their leadership model, so that whatever part the user engaged with, an understanding of this particular view of leadership would be the learning outcome. Even learning to operate the user interface of the simulator would give a better understanding of leadership.

However, although there are examples of the development of training simulators, and although methodologies for specifying training simulator requirements have been formulated, there is a void between these and the user-centred methods. An approach to User-Centred Design for training simulators is needed, where the particularities of simulator development are accounted for.

3 Methods

Motivated by this need for a User-Centred Design approach to the development of training simulators, an exploratory study was conducted within a project for developing a simulator for negotiation training. The project was initiated by an associate professor at the University of Oslo, and carried out without the involvement of any external stakeholders. The design team was composed of five master's students in psychology, all dedicating one year to this project while writing their master's thesis on its different aspects. Some of the members had prior experience with development processes in general and the User-Centred Design process in particular, while others had little or no such experience. A laboratory at the Department of Psychology in Oslo had been assigned to the project, in which the development activities took place, with the exception of one activity that was carried out in a Usability laboratory located in the centre of Oslo. All five members of the design team were present at all times and collaborated on the majority of the development tasks.

The aim of this exploratory study was mainly to generate ideas for the application of User-Centred Design methods to the development of training simulators, and the project served as a testbed where relevant user-centred methods were applied and their use and effects continuously logged and evaluated. As this study was conducted by one of the five members on the design team, in the role of a participating observer, methods and techniques could be chosen with the criterion that they should be suited for the design process as well as have a potential for generating inputs to the ongoing research, and first-hand experience of their application could be collected as data and form the basis for generating new ideas (Meyer, 1992). Throughout the project all relevant activities were logged, with descriptions of what was done, the group's reasoning behind the process and techniques, observations from the work, and online reflections on behalf of the observer. The theoretical structuring was postponed until the final analysis, in accordance with the suggestion made by Flick (2002).

4 Results

The results presented here are based on observations and reflections recorded during the design process. The presentation is structured according to three of the main activities in User-Centred Design: analysis, design, and user testing. Only the parts relevant to this study are presented.

4.1 Analysing a field of practice as a context of use

Two investigations of negotiation as a context of use were conducted during the project period. At the beginning of the first iteration, professionals with negotiation experience were interviewed to gain an overview of the area and details about the task of negotiating. In addition to general questions, the interviewees were asked to describe the sequence of events in a (preferably) recent negotiation. In the second iteration, professionals were observed while they negotiated in sessions arranged by the design team. Afterwards the participants were shown a videotape of the negotiation, and were asked to comment on their actions and explain, as much as possible, what they did and why.

The data from the first iteration were analysed with techniques from Contextual Design (Beyer & Holtzblatt, 1998), through several sessions where the design team first interpreted the data from each interview, resulting in a list of key statements and models of how the interviewees perceived and practiced negotiation. The models were then consolidated as generalised models spanning the individual cases, and the statements were organised hierarchically in an Affinity Diagram (for a more detailed description, see Beyer & Holtzblatt, 1998). Only two CD work models were considered relevant for the collected data. The Sequence Model presented a generalised description of a negotiation in phases, steps and intentions, displaying the common structure across interviews. The Cultural Model (not included here) gave an overview of the groups and people involved in a negotiation, and their influences on and attitudes toward each other. Due to diverging configurations of people, groups, and roles from negotiation to negotiation, the common structure across interviews was not as apparent for this model as it was for the Sequence Model. Data from the second iteration were analysed in a similar manner, extending and refining the Affinity Diagram.

Together with the two models, the Affinity Diagram gave the design team a broad impression of how negotiation is practised, and pointed to central aspects. It was the design team's experience that this gave a much closer understanding of the interviews, as the data was evaluated several times, both individually and as a group. This was especially true with the Affinity Diagram, as each statement was glued to a Post-it, placed on the wall, and then grouped and regrouped several times until an overall structure emerged. The physical presentation and the size of the diagram made it possible to analyse the data together as a team. Individual members also claimed to have had their first impressions of the tendencies in the data altered and expanded through the interpretation and consolidation, resulting in a different presentation of the interviews than the group's initial more intuition-based experience of commonalities across the interviews.

Furthermore, the models, especially the Sequence Model, and the Affinity Diagram were important inputs to the creative process, both as a common point of reference for the design team and as aids for entering into the right mindset – the team members would revisit the models and the diagram at the beginning of design sessions. Qualitative differences were observed between discussions on topics included in the models or the Affinity Diagram and on those that were not. As an example, an idea regarding the use of a mediator in the simulated negotiation came up during the design process. There was actually a lot of data on mediation, but this had not been included in the first round of analysis due to time constraints and a decision to narrow the focus of the analysis. While evaluating this idea, the team members made arguments like 'but he said in the interview ...' and 'she always used to ...', referring to the negotiators that had been interviewed. For other ideas, references were made to the diagram and the models, and not to specific interviews. Thus the models and the diagram seemed to have made the design team less focused on the specifics of individual cases, and instead induced a focus on the general tendencies in the data.

However, the design team voiced a concern that the original meaning of the interviewees might not be preserved through all stages of the analysis process, from the interview, through the interpretation into the individual models and statements, and finally to the consolidated models and the Affinity Diagram. The diagram also seemed to hide many specifics, especially on topics mentioned by only one or two of the interviewees. For example some specific negotiation strategies that could have been important to include in the simulator, did not show up on the labels in the final diagram. They were present below more general labels, grouped with other related statements, but the specific strategies were lost in the hierarchy.

4.2 Designing a simulator-based training programme from user data

Inspired by Contextual Design (Beyer & Holtzblatt, 1998) and Cognitive Work Analysis (Vicente, 1999) the aim of the development process was to design a training programme in negotiation, supported and facilitated by the simulator, as opposed to designing a simulator only. With the Sequence Model, the Culture Model, and the Affinity Diagram as input, the design team started the creative process by brainstorming several different visions of how a training course with the simulator would look like. Based on considerations of positive and negative aspects of each, two common visions were created from the most promising ideas, one for the training programme and one for the simulator (not included here). Usually in CD, only one common vision is created for the entire design of a new way of working. While the simulator vision was really a part of the training programme vision in this case, a split was made to render the design decisions clearer.

In the spirit of Contextual Design, the visions were then explored in more detail by drawing storyboards of the interaction between a user and the simulator in an imagined scenario, as a series of cartoon-like images. This allowed an explication of parts of the design without having to make a prototype of it, and to make decisions on a more detailed level than is possible in a vision. Storyboards are intentionally kept on a conceptual level, not going into technical details, so the design discussions initialised by this technique were mainly concerned with the user's interaction with the system and the virtual opponent in the simulation, and not with the inner workings of the simulator. These sessions were focused on what parts of the simulation the user would need to control in order to have the necessary degrees of freedom to practice negotiation in a way that would seem sufficiently realistic and involve issues relevant to learning this skill. This was a constructive process that both generated new ideas and uncovered issues that had to be solved for the design to work.

As an example of an issue uncovered through storyboards, the vision for the simulator included an idea motivated by data on the importance of taking the opponent's perspective in a negotiation, were the user would get the opportunity to actually have a look at the opponent's 'cards', so to speak, by switching view either during the negotiation or immediately afterwards. In this way the user could compare his or her impression of the opponent's position and priorities with the actual situation in the simulation, and reflect on any discrepancies. However, when this came up in the storyboard, some team members began to question whether this feature would really teach a user to consider the opponent's perspective. Instead, it could make him or her more focused on securing a good deal, by allowing a comparison of the end-result of a negotiation with how much further the opponent would have been willing to go. This could lead to a view of negotiation as an effort to get as much as possible from the other party – a view the design team would not necessarily advocate. The conclusion of the discussion there and then was that the initial idea of switching view had come up because it was possible, rather than being a good way of facilitating the desired learning objective. Through the focus of the storyboard technique the design team were able to discover this issue and find alternative and more suited ways of focusing on perspective taking.

4.3 Testing simulator-based training with users

Based on the design from the first storyboard session, a prototype of the simulator was created and tested with users, in accordance with User-Centred Design approaches in general (Beyer & Holtzblatt, 1998; Faulkner, 2000; ISO 13407, 1999; Kuniavsky, 2003; Mayhew, 1999; Nielsen, 1993). However, the task of recruiting participants for this test raised the question of who would be relevant users for the training simulator. On the one hand, the end-users of the simulator would be persons seeking training in negotiation, but as a consequence of this characterisation, they would not be able to evaluate whether the training is relevant to the task of learning to negotiate and whether the simulation is a satisfying representation of negotiation. On the other hand, testing on experts of negotiation could provide inputs on the realism and accuracy of the simulation, but by itself this would not provide any indication that the simulator would actually facilitate learning (Alessi, 1988; Farmer et al., 1999).

At this stage in the development, the main focus was to test the general interaction between the user and the simulator and whether it was experienced as a conversation with the virtual character presented in the prototype. Thus, for this first Usability test the participants' level of experience with negotiation was not crucial, and available students at the University were recruited. The test provided valuable input on the design and the general design concept, but the question of who would be relevant users for future tests remained unanswered.

Also, at this point, it became clear that this user-centred process had not really provided any training objectives for the evaluation of the design. A lot of analysed data on negotiation had been gathered to drive the creative design process, and these data were in themselves a set of requirements for the design, but from a Usability perspective only – not from a pedagogical perspective. In other words, the user-centred analysis had resulted in input on what to include in the simulator in order to design an interface for a negotiation between a human user and simulated characters, but not for the purpose of providing training in the skills involved in negotiation. Hence, the user test provided the design team with input on the Usability of the simulator, but not on training aspects.

5 Discussion

This paper has been concerned with the application of User-Centred Design methods to the development of a training simulator. Most of the techniques in the study were adopted from Contextual Design (Beyer & Holtzblatt, 1998) or User-Centred Design in general. While the results are inconclusive in the sense that they do not span a complete development process from first analysis to finished product, the aim of the study was to generate ideas for User-Centred Design of a training simulator, and several tendencies have been discovered.

The development project has served as a demonstration of how a field of practice, in this case negotiation, could be analysed through user-centred methods for the purpose of designing a simulation of it. In this regard, the conclusion of the study is that the Sequence Model, the Cultural Model, and the Affinity Diagram from Contextual Design, including the techniques for generating them, may provide the design team with a broad understanding of the field and a point of reference for the creative process of designing the training programme as well as the interaction with the simulator. These techniques directed the focus in the creative process to the commonalities across cases, rather than personal opinions, and specific and possibly contradictory cases.

That said, the experiences from this study suggest that the Affinity Diagram and the models will not provide the necessary level of detail and accuracy to inform the design of rules for the inner workings of the simulator, such as the artificial intelligence needed to control the behaviour of the virtual characters (Aldrich, 2004). The many steps of interpretation from data collection to the final, consolidated models and diagram may transform the original meaning of the data and, in the worst case, lead to skewed or faulty models of the field of practice. Any loss of meaning in the presented case study may be due to the design team's way of applying the techniques, but on the other hand, the data gathering and analysis techniques in CD are based on building understanding in the members of the design team, and in several of the steps this understanding is the main carrier of data to the next step. Thus, a loss of meaning or a bias may be expected by any design team using these techniques. Moreover, the hierarchical structure of the Affinity Diagram in this case revealed few specifics and was mostly general in nature. The generality in the diagram may stem from a lack of specificity in the original data, as the first round of analysis in this case was based on interviews and not on observations of negotiation. People's own accounts of what they do are generally less detailed than their actions in a real situation (Suchman, 1987). It may also be that a topic and a field as large and complex as negotiation would need more than three levels of categories in an Affinity Diagram to reveal any details. A third possibility is that generality may in fact be a necessary property of a diagram generated through a consolidation process. Regardless of the source of this generality, the study indicates that the specifics needed for the rules of the simulator may not be provided through the Contextual Design techniques. Note that this is not intended as a critique of CD per se, as its techniques were never developed to meet the criteria set by this discussion; the intention is rather to clarify its possibilities for application in the design of a training simulator.

The principle of Contextual Design and the formative approaches in general, to redesign work rather than just design a tool to support it, may not at first seem to apply to the development of a training simulator. Instead of redesigning the analysed task, it should be simulated with a certain level of fidelity to allow the skills acquired to be transferred to a real situation (Alessi, 1988; Farmer et al., 1999). To exemplify, a redesign of the negotiation activity could involve a computer application that would take both parties' interests and priorities as input and calculate the most advantageous outcome, but as this is not how negotiations are conducted, this is not what should be designed into the simulation either. However, while the field of practice should be simulated as it is analysed, the training activity could be designed or redesigned – supported by the simulator. This was demonstrated in the presented study, and gave the design team a different perspective on the design process and the issues related to the actual simulator; an example taken from the results is the idea of letting the user switch focus and see the negotiation from the opponent's perspective. The idea had come up because it was a possible feature, without an evaluation against the intended training outcome, yet through a design process with focus on the entire training programme and not just the supporting artefact, the design team was able to discover this confusion. The idea may also be attributed to a more conventional User-Centred Design thinking, as it could have been a plausible feature had the goal been to redesign negotiation, rather than simulate it. This supports the claim that a clarification of the involved systems and their differing roles is needed.

Even though the design team had a focus on designing the training activity, they presented their overall design as two distinct, but related parts: the training programme and the simulator system. One can only speculate as to whether this specific division was triggered by a real need of the process or by an a priori distinction on behalf of the team members, but the distinction was nevertheless experienced as both meaningful and necessary by the design team. Thus, while the process gained from the emphasis on designing the entire training activity through a Training Vision, a separate focus on the simulator system, with a separate Simulator Vision, was also valuable in this process.

A final challenge stemming from the User-Centred Design approach was discovered by the question of whom to test the simulator and the training design with – experts that can verify a correct simulation of negotiation or end-users with little negotiation experience, that must be able to use the simulator in order to learn from it? A first intuition on this dilemma would be to simply test the simulator on both of these groups, but a usable simulator with high fidelity does not necessarily ensure that novices will actually learn the skills relevant for negotiation in a real setting (Alessi, 1988; Farmer et al., 1999; Squires & Preece, 1996). Furthermore, as the user testing technique is based on the definition of Usability (Faulkner, 2000; Nielsen, 1993), its application will inevitably focus on whether the user is able to operate and understand the simulator instead of on the training outcomes, and again the effect may be that the team designs a simulator to support negotiation between the user and virtual characters rather than to support training. It is essential to make sure that the user will be able to operate the simulator and to ensure a certain degree of fidelity, but these should not be the only aspects of the design–test iterations. The evaluation of the training activity and the simulator should focus on training objectives and pedagogical issues as well. In the extension of this, a method for developing training simulators should also include techniques for specifying a set of learning objectives to evaluate the design against, which the User-Centred Design methods obviously do not.

As a summary of this discussion, techniques from Contextual Design may be used for designing the training programme and the interface of the simulator, but more detailed and accurate techniques are needed to inform the design of the inner workings of the simulator; a perspective where one designs the entire training programme should be adopted in order to give the design team a focus on supporting training, not only simulating an activity, although it may be useful to approach the overall design as a Training Vision and a Simulator Vision. Furthermore, while Usability tests are necessary to ensure that the users are able to operate and understand the simulator, and fidelity checks with domain experts may be important to promote transferability of the acquired skills, the techniques for defining and assessing the training objectives with users are needed as well.

6 General discussion

This paper has so far pointed to some requirements for User-Centred Design methods for training simulators. First of all, guidance on the different roles of the systems involved in the development is needed. Second, techniques must be included for providing data for the creative design of the training programme and the simulator interface, as well as for the rules and details of the inner workings of the simulation, based on the principle of designing a simulator-supported training activity, not only a simulator. Third, the method must support the design team in defining training objectives for the simulator and overall training activity, and supply techniques for assessing whether the design is capable of fulfilling these, as well as evaluating the Usability of the simulator with learners and the fidelity of the simulation with domain experts. In the following, a possible approach for such a method is suggested.

6.1 The different roles of the involved systems

To start with the first requirement, a clear depiction of the systems involved in the development process is needed. Two systems have been recognised from the start: the system to be supported – the training activity, and the system to be simulated – the field of practice to provide training in. In addition, a third system was mentioned in the previous discussion – the simulator system itself – given its natural and distinct role in the studied design process. These systems are also found in the three main phases of requirements specifications in the MASTER project (Farmer et al., 1999), which give separate focus to the field of practice, the training activity, and the simulator. The relationships between these systems are given in the earlier description of the system to be supported and the system to be simulated as activities, namely that (a) the field of practice is represented in the training activity through the simulator, (b) the skills acquired in the training activity should be transferable to the field of practice, and (c) the user will simultaneously be engaged in the activity of training and the simulated activity from the field of practice. To elaborate on the role of the simulator system, Kuutti (1996, p. 34) mentions the possibility of technology to provide a 'window' into an activity, and this is how technology is used in the case of a training simulator, although possibly in a more direct sense than intended by Kuutti. The role of the simulator system is thus to provide the user with a window through which he or she may engage with the field of practice as a part of the training activity. Figure 8.2 shows a model of the three systems, with their three types of relations represented as arrows.

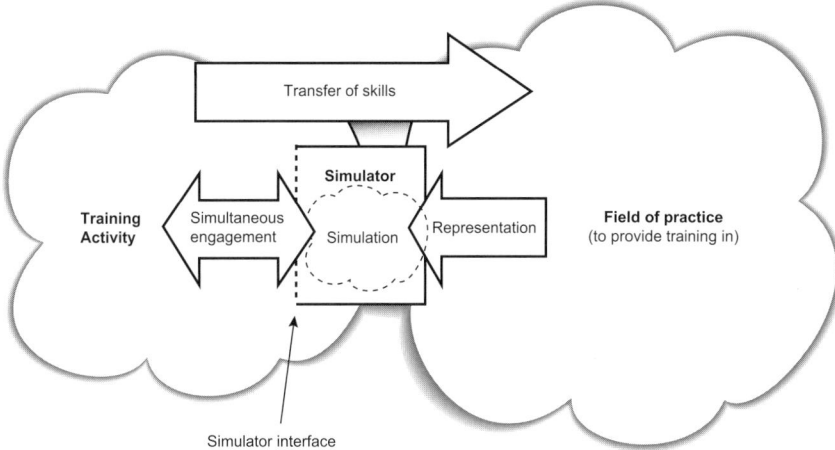

Figure 8.2. A model of the three systems involved in the development of a training simulator, and their relations.

From this, it follows that the purpose of the design process is to enable the relations between the three involved systems. Through an analysis of the field of practice, a representation is designed into the simulator system allowing the user to engage in the field of practice as a part of the training activity, in a way that leads to the acquisition of skills that may be transferred back to the field of practice. The training activity and the simulator system are the ones to be developed through the design process, indicated by the thicker lines around these systems in Figure 8.2, where the simulator will have status as a tool for the training activity.

6.2 Informing the design of training activity, simulator interface, and simulation

As for the second requirement, the techniques of Contextual Design (Beyer & Holtzblatt, 1998) may very well be used for the design of the overall training programme and the simulator interface, as demonstrated in this case study. The field of practice would be analysed and presented in an Affinity Diagram and any of the five models relevant to the particular field of practice. On this basis, the design team may lay the foundation for the training activity and the simulator interface by creating a vision for each, and elaborating through storyboards. The emphasis of this process should be on the training activity and the acquisition of skills that can be transferred to the field of practice, and this should dictate the design choices for the simulator, and not the other way around. Also, user-centred analysis of an existing training activity could be useful to inform the design of both the training activity and the simulator, in addition to their continuous evaluation through Usability tests and other relevant techniques.

When choosing techniques to provide input to the inner workings of the simulator, Cognitive Work Analysis (Vicente, 1999) seems to be a strong candidate, for several reasons. First, the method's heritage from Cognitive Engineering, with a focus on sociotechnical systems rather than artefacts, complies with the requirement for designing the training activity and not only the simulator. Second, its modelling tools for finding the intrinsic constraints of the different layers of analysis are more detailed and scientifically based than the creative techniques of CD, giving reason to expect that CWA may provide the level of detail necessary for specifying the rules governing the simulation. Third, and most interesting, learning is at the centre of CWA, as its analysis starts with the constraints of the environment rather than the cognition of the user, in order to provide the user with a more realistic mental model. This forming of a mental model in the user is in essence learning. The intrinsic constraints of the activity in the field of practice may be analysed and represented as the space of action possibilities in the simulation, and by exploring these constraints, the user could develop a mental model of the field of practice that may be applicable in a real situation. Furthermore, the idea of defining a constrained space for possible actions, as opposed to designing a set sequence of events through the field of practice, is compatible with Aldrich's (2004) principle of providing the user with cyclic and open-ended content. Instead of following a predefined trajectory through the content, the user will be free to explore, apply different strategies, and test the boundaries of the simulation in order to build an understanding of the field of practice in a constructionist way of learning.

That said, CWA and its modelling tools are developed in the setting of more tangible and established sociotechnical systems, such as process control systems (Vicente, 1999), and not as a tool for modelling social activities such as negotiation. However, its techniques have been used to analyse activities as different as design processes (Rasmussen, 1990), the game of chess (Vicente & Wang, 1998), interdisciplinary interactions (Burns & Vicente, 1995), and medical examinations (Hajdukiewicz, Vicente, Doyle, Milgram, & Burns, 2001). Vicente (1999, p. 304) also claims that CWA is not limited to 'domains where workers are controlling a physical system whose behavior is governed by laws of physics that can be described by equations'. Based on this, it may very well be possible to analyse negotiation as a Work Domain in CWA, with the goals of the activity as Control Tasks, and with several Strategies for reaching each goal; though a further explication of this is beyond the scope of this paper. The Social-Organisational task allocation and co-operation has already been mentioned in the description of the field of practice in terms of Activity Theory (Kuutti, 1996), as the division of labour, and the fifth conceptual distinction, Worker Competencies, will be returned to later in this discussion.

With the techniques from Contextual Design to inform the design of the training programme and the simulator interface, and the models from Cognitive Work Analysis as specifications for the simulator, a link between the two is still missing. Both will provide an analysis of the field of practice, but they represent two distinct approaches to this analysis, with different models and diagrams. A solution may be found in Vicente's (1999, p. 110) description of the formative approach as based on conceptual distinctions with 'direct and obvious implications for design'. The five models of Cognitive Work Analysis will inform the design of the simulation in much the same way as described by Vicente, although with emphasis on simulating rather than redesigning the analysed context, and on giving realistic possibilities for action

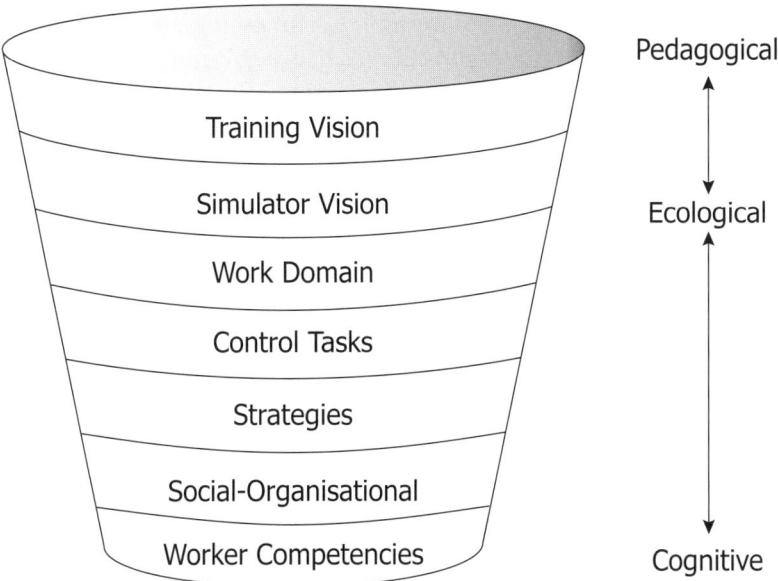

Figure 8.3 The layers of analysis for a training simulator (adapted from Vicente 1999).

to both virtual characters and human users, rather than distributing responsibilities between workers and automation. Contextual Design, on the other hand, will be applied with implications for the training programme and the simulator interface, through the two conceptual distinctions presented by the Training Vision and the Simulator Vision. The former poses constraints through overall decisions on which parts of the field to supply training in, while the latter represents the boundaries of the simulator as a window into the field of practice.

In Figure 8.3, the Training Vision and the Simulator Vision are depicted as two layers of analysis above the layers from Cognitive Work Analysis, representing a transition from pedagogical considerations to the ecological considerations of CWA. This order is in accordance with one of the main tenets of the MASTER approach, that requirements for the simulator should come from training considerations rather than the details of the system to be simulated (Farmer et al., 1999). Also, the notion of the simulator as a window into the field of practice would imply that the boundaries of the window need to be defined before the analysis of what the user will see through it. Vicente (1999, p. 171) recommends 'defining the scope of the analysis by drawing a boundary to delimit the work domain of interest' before conducting a Work Domain Analysis, which is precisely what is done through the visioning process for the training programme and the simulator.

An objection to the approach described here may be that a vision in Contextual Design is not a model of intrinsic constraints from the field of practice, as are the models in CWA. While this is correct, there is no direct link from the intrinsic work constraints in a field of practice to a set of useful constraints for a training programme, meaning that the definition of these constraints will necessarily have to be based on decisions made by the design team. Furthermore, similar decisions are made through the analysis in CWA. For example, an analysis may include decisions on the distribution of tasks and responsibilities between human workers and automation in the new sociotechnical system to be designed. It should therefore be equally legitimate to constrain the analysis of the field of practice based on the overall decisions for the training programme and the simulator interface, especially since these are grounded in data from the field of practice as well.

6.3 Defining and evaluating training objectives

With regard to the requirement for a technique for defining training objectives, the MASTER approach (Farmer et al., 1999) may provide some insight. In the MASTER approach, the discrepancies between required and available skills form the basis for defining training objectives. This is the end result of the Training Needs Analysis, which progresses from an analysis of the overall activity and the tasks performed within this activity, to an analysis of the skills required and the skills currently possessed by the people to be trained. This progress bears resemblance to the transition from environment to cognition in CWA analysis, and the modelling tool for Worker Competencies – the last level of analysis for intrinsic work constraints in CWA – is not only intended for analysing the constraints of the users' cognitive capacities and the skills needed for satisfactory performance but may also provide an analysis of the current competencies of users as well (Vicente, 1999). Thus, by analysing the required and available worker competencies, CWA is not only promising as a technique for informing the design of the inner workings of the simulator, but may also be used as a technique for defining training objectives for the simulator and the overall training activity.

Then the only remaining requirement is techniques for evaluating the training outcomes and their relation with Usability testing of the interface and fidelity checks with domain experts. The description of specific techniques for evaluation of training outcomes is outside the scope of this paper, but it may be argued that Usability and educational aspects should be seen as interacting issues rather than two distinct dimensions (Squires & Preece, 1996). Attempts have been made at integrating the principles of Usability with modern learning principles in order to perform expert evaluations on both aspects (Squires & Preece, 1999). Also, participants may be observed and interviewed as part of the test with the training objectives in mind, to find indications of learning or the lack thereof, as input to the design process (Rappin, Guzdial, Realff, & Ludovice, 1997). On the other hand, it has been argued that usability issues should be addressed first, since the ability to use a system is a prerequisite for learning from it (Quinn, 2005), but what parts of the simulator to make usable should nevertheless be seen in relation to the task of learning, as struggling with a task may actually be necessary for learning (Mayes & Fowler, 1999). This discussion is in line with an ongoing discussion in the educational computing literature (Squires, 1999), and no clear solution to the dilemma exists to date (Ardito et al., 2006).

When it comes to conducting fidelity checks with domain experts, applying a Usability approach to test whether the experts are able to negotiate with the virtual characters in the simulation may actually be valuable. If experienced negotiators consider the simulated negotiations to be realistic, the chances are higher that the correct constraints on the space of possible actions have been identified, and skills acquired through training with the simulator may be transferable to real situations. However, this testing should be done with some care, as this perspective of supporting negotiation should not be adopted for any other part of the design process. Also, a faithful simulation is not necessarily related to learning, and deciding to what extent and in which way fidelity is a goal for the design process is essential (Alessi, 1988; Farmer et al., 1999). While tests with learners may integrate the Usability and educational perspectives, tests with domain experts from a fidelity perspective have to be carried out separately, as different participants are involved. Nevertheless, these latter tests should be seen in close relation with the former, to ensure that they serve the overall goal of training and not that of supporting negotiation.

7 Conclusions

Based on the experiences from a User-Centred Design process for developing a simulator for negotiation training, a user-centred approach to the development of simulator-based training has been proposed, founded primarily on Contextual Design (CD) for the design of the overall training activity and the simulator interface, and Cognitive Work Analysis (CWA) for the rules and details of the inner workings of the simulator. Through a formative approach where the field of practice is analysed in order to define constraints for the user's actions in the system, a Training Vision and a Simulator Vision are created through techniques from CD, to represent the constraints of the training activity and the simulator system respectively. Within these constraints, the field of practice is analysed through the five conceptual distinctions from CWA in order to inform the design of the simulation. This leads to a space of possible actions in which the user may explore the simulated field of practice to build an understanding and a mental model that may be transferred to real situations. The last conceptual distinction of CWA, Worker Competencies, may also provide training objectives for the simulator by analysing the discrepancies between the required skills for the field of practice and the potential users' existing skills. Based on these training objectives, the design of both the training activity and the simulator may be tested on potential users and domain experts, with an integrated perspective on training, Usability, and simulation fidelity.

Research is still needed to evaluate whether Cognitive Work Analysis may be applied to the analysis of a social reality, such as negotiation, and whether the Cognitive Design visions may serve as constraints on top of CWA's layers of analysis in an actual design process. Also, more work is needed in the pursuit of integrated techniques for evaluating the Usability, fidelity, and pedagogical issues of simulator designs. Hopefully, however, the approach suggested in this paper will provide designers with better methods for developing training simulators that will facilitate learning of useful and transferable skills.

References

Aldrich, C. (2004). *Simulation and the future of learning*. San Fransisco: Pfeiffer.

Alessi, S. M. (1988). Fidelity in the design of instructional simulations. *Journal of Computer-Based Instruction, 15*(2), 40–47.

Ardito, C., Costabile, M. F., De Marsico, M., Lanzilotti, R., Levialdi, S., Roselli, T., & Rossano, V. (2006). An approach to usability evaluation of e-learning applications. *Universal Access in the Information Society, 4,* 270–283.

Bertalanffy, L. von (1972). The history and status of General Systems Theory. *The Academy of Management Journal, 15*(4), 407–426.

Beyer, H. & Holtzblatt, K. (1998). *Contextual Design: Defining customer-centered systems.* San Francisco: Morgan Kaufmann Publishers, Inc.

Burns, C. M. & Vicente, K. J. (1995). A framework for describing and understanding interdisciplinary interactions in design. In *Proceedings of DIS 195: Symposium on Designing Interactive Systems* (pp. 97–103). NY: ACM Press.

Carroll, J. M. & Campbell, R.L. (1989). Artifacts as psychological theories: The case of human-computer interaction. *Behaviour and Information Technology, 8*(4), 247–256.

Farmer, E., Rooij, J. van, Riemersma, J., Jorna, P., & Moraal, J. (1999). *Handbook of simulator-based training*. U.K.: Ashgate.

Faulkner, X. (2000). *Usability Engineering*. U.K.: Palgrave.

Flick, U. (2002). *An Introduction to Qualitative Research, 2nd ed.* London: Sage Publications.

Gould, J. D. & Lewis, C. (1985). Designing for Usability: Key principles and what designers think. *Communications of the ACM, 28*(3), 300–311.

Gredler, M. E. (2004). Games and simulations and their relationships to learning. In D. H. Jonassen (Ed.), *Handbook of research on educational communications and technology* (pp. 571-581). Mahwah, NJ: Lawrence Erlbaum.

Hajdukiewicz, J. R., Vicente, K. J., Doyle, D. J., Milgram, P., & Burns, C. M. (2001). Modeling a medical environment: an ontology for integrated medical informatics design. *International Journal of Medical Informatics, 62,* 79–99.

Hollnagel, E. (1998). Comments on 'Conception of the cognitive engineering design problem' by John Dowell and John Long. *Ergonomics, 41*(2), 160–162.

ISO 13407 (1999). *Human-centered design processes for interactive systems.*

Kafai, Y. & Resnick, M. (1996). *Constructionism in practice: Designing, thinking, and learning in a digital world.* N.J.: Lawrence Erlbaum Associates, Inc.

Karat, J. (1996). User Centered Design: Quality or Quackery? *Interactions, 11*(4), 19–20.

Koschmann, T. D. (1996). Paradigm shifts and instructional technology: An introduction. In T. D. Koschmann (Ed.), *CSCL: Theory and practice of an emerging paradigm* (pp. 1–23). New Jersey: Lawrence Erlbaum Associates Publishers.

Kuniavsky, M. (2003). *Observing the user experience: A practitioners guide to user research.* SF: Morgan Kaufmann Publishers.

Kuutti, Kari (1996). Activity Theory as a potential framework for human–computer interaction research. In B. A. Nardi (Ed.), *Context and Consciousness: Activity Theory and human–computer interaction* (pp. 17–44). MA: Massachusetts Institute of Technology.

Mayes, J. T., & Fowler, C. J. (1999). Learning technology and usability: a framework for understanding courseware. *Interacting with Computers 11,* 485 – 497.

Mayhew, D. J. (1999). *The Usability Engineering Lifecycle: A practitioner's handbook for user interface design.* SF: Morgan Kaufmann Publishers.

Meyer, M. A. (1992). How to apply the anthropological technique of participant observation to knowledge acquisition. *IEEE Transactions on Systems, Man, and Cybernetics, 22*(5), 983–991.

Nielsen, J. (1993). *Usability engineering.* CA: Academic Press.

Norman, D. A. (1986). Cognitive Engineering. In D. A. Norman & S. W. Draper (Eds.), *User centred system design: New perspectives on human-computer interaction*. New Jersey: Lawrence Erlbaum Associates Publishers.

Norman, D. A. (2002). *The design of everyday things*. NY: Basic Books.

Quinn, C. N. (2005). *Engaging learning: Designing e-learning simulation games*. SF: Pfeiffer.

Rappin, N., Guzdial, M., Realff, M., & Ludovice, P. (1997). Balancing usability and learning in an interface. In *Proceedings of the ACM Human Factors in Computing Systems Conference (CHI'97)* (pp. 479–486). NY: ACM Press.

Rasmussen, J. (1990). A model for the design of computer integrated manufacturing systems: Identification of information requirements of decision makers. *International Journal of Industrial Ergonomics, 5*, 5–16.

Salas, E., Bowers, C. A., & Rhodenizer, L. (1998). It is not how much you have but how you use it: Toward a rational use of simulation to support aviation training. *International Journal of Aviation Psychology, 8*(3), 197–208.

Sommerville, I. (2004). *Software engineering*. London: Addison Wesley.

Squires, D. (1999). Usability and Educational Software Design: Special Issue of Interacting with Computers. *Interacting with Computers, 11*, 463–466.

Squires, D., & Preece, J. (1996). Usability and learning: Evaluation the potential of educational software. *Computers Educ., 27*(1), 15–22.

Squires, D., & Preece, J. (1999). Predicting quality in educational software: Evaluating for learning, usability and the synergy between them. *Interacting with Computers, 11*, 467–483.

Suchman, L. A. (1987). *Plans and situated actions: The problem of human–machine communication*. Cambridge University Press.

Vicente, K. J. (1998). An evolutionary perspective on the growth of cognitive engineering: the Risø genotype. *Ergonomics, 41*(2), 156–159.

Vicente, K. J. (1999). *Cognitive Work Analysis: Toward safe, productive, and healthy computer-based work*. NJ: Lawrence Erlbaum Associates Publishers.

Vicente, K. J. (2004). *The human factor: Revolutionizing the way people live with technology*. NY: Routledge.

Vicente, K. J. & Wang, J. H. (1998). An ecological theory of expertise effects in memory recall. *Psychological Review, 105*(1), 33–57.

Woods, D. D. & Hollnagel, E. (2006). *Joint cognitive systems: Patterns in Cognitive Systems Engineering*. FL: Taylor & Francis Group, LLC.

9

A Video-Based Phenomenological Method for Evaluation of Driving Experience in Staged or Simulated Environments

Kjell Ivar Øvergård

This paper explores the possibilities of adopting qualitative methods in the evaluation of traffic safety measures. To date, few published studies have employed qualitative methods and the used methods today mainly generate quantitative data. These data are external to the experience of the driver. A qualitative method based upon the phenomenological philosophy of Husserl and Merleau-Ponty is presented to identify the experiential aspects of driver behaviour. This research design can be used in conjunction with psychological experiments and can uncover how quantitative and qualitative levels of data relate and cross-fertilize each other.

1 Introduction

Transportation and traffic safety research is a diverse field where different scientific approaches are involved, among which are engineering, psychology, human factors and logistics (Groeger & Rothengatter, 1998). All these approaches work to effectively reduce risk related to transportation and traffic behaviour, and in this respect they have been successful to some extent. The risk of being killed or harmed in traffic accidents has been reduced today compared to ten, twenty or thirty years ago. This is mostly due to better passive safety in cars, due to the introduction of safety belts, airbags, and collision deformation zones in cars, for example. Further, changes in the road environment have contributed to reducing the consequences of accidents (Rothengatter, 2002). Even though measures have been taken to reduce the number of traffic accidents and the consequences of them, accidents still happen. In Norway, 242 persons were killed, 940 persons were seriously injured, and 8866 persons sustained minor injuries in a total of 7925 reported traffic accidents in 2006 (Statistisk Sentralbyrå, 2006). This shows that it is still a challenge to reduce this number of accidents and fatalities.

Different levels of analysis are used to identify the causes of traffic accidents. The layout of the road and its surroundings (engineering factors), the design, age and condition of the vehicle, and psychological or human factors of the drivers or other road users are all features reckoned to play a part in accidents and contribute to their consequences. Estimates of the causes of accidents reveal that human error probably contributes to about 90% of all traffic accidents. These causes can be a lack of precautions regarding speed choice, misinterpreting road conditions, and the mental and/or physical status of the driver, due to deprivation, drinking and driving (McKenna, 1993; Hills, 1980). If the estimates are correct, great gains could be made in relation to affecting the driver and his/her choice of driver behaviour. For example, researchers could work towards influencing drivers' attitudes, speed choice, and taking precautions – and thus induce safer behaviour in the traffic environment. This research strategy has been pursued, but the applied results have been somewhat less effective than expected. Some researchers argue that the potential for influencing drivers has not yet been fulfilled (Rothengatter, 2002). Since the current approaches have not had as large effects as could be expected, it is possible that new approaches and method should be used to explore the driver experience in order to guide our understanding of why drivers act as they do – thus enabling us to target interventions more precisely.

1.1 The aim of the paper

The paper explores the possibilities of adopting qualitative methods in the evaluation of traffic safety measures. It will also show how qualitative methods can help improve traffic safety measures in relation to the understanding of driver behaviour. The line of reasoning in this paper builds on phenomenology as a starting point for a new way of evaluating and conducting traffic safety research. Other researchers have conducted phenomenological analyses of driving (e.g. van Lennep, 1987), but these have not been directly related to traffic safety. The suggested method in this paper makes it possible to show how the subjective and experiential world relates to changes in the external world.

This research strategy breaks with the leading research paradigms in traffic safety research in that it accepts subjective or experiential phenomena as valid psychological data. This paradigm discards the Cartesian mind-body dualism inherent in positivism (Giorgi, 1970), and it does so by reference to the key phenomenological assumption suggesting that a person experiences the world through her/his body (Merleau-Ponty, 1962). This means that in analyses it cannot be useful to differentiate between what is inside the mind and what is inside the body. A phenomenological perspective on human thought and behaviour regards the mind as the product of the embodied interaction of the human organism and its habitat (Lakoff & Johnson, 1999). Before the phenomenological paradigm is described, a summary presentation of the current and classical perspectives used in traffic safety research is offered in order to clarify how phenomenology differs and the perceived anomalies it aims to relieve.

2 The Extended Use of Quantitative Data within Traffic Safety Research

Traffic safety research is a highly transdisciplinary field of research (Groeger & Rothengatter, 1998). However, some traditions are more salient than others, and three distinct approaches to traffic research will be summarized in the following: the engineering approach, traffic psychology approach, and psychophysiology approach.

2.1 The engineering approach

Much of the research within traffic safety has been carried out using an experimental approach closely related to the positivistic research paradigm. Diverse forms of controlled experiments have been the favoured method in research programmes. Experimental control of the variables included in the study is seen as the norm for the construction of valid scientific knowledge. This is also true for the engineering approach to traffic safety. Controlled experiments are looked upon as the norm by which driver behaviour should be studied. In terms of the philosophy of science, scientific objectivity emphasizes that data should be uncontaminated by the attitude and opinion of the researcher. This stance is evident in the experimental approach of traffic safety researchers, although they may be clear on the fact that pure scientific objectivity is not possible (Elvik, Vaa & Østvik, 1989).

The engineering approach peaked in popularity in 1972. Typical interventions often involved altering the physical design of roads and surroundings to reduce accident risk and the consequences of accidents, or altering the design of cars (Rothengatter, 2002). Measures such as separating driving lanes in opposite directions, increasing lane width, altering speed limits, and introducing seat belts in cars have been very important in reducing traffic accidents and fatalities. This approach is one of the most efficient in reducing accidents and the loss of human lives in traffic accidents (Rothengatter, 2002).

In the engineering approach, the ways of collecting data have often been to introduce a traffic safety measure, and then monitor accident frequencies, fatalities or driving speed change as a function of this measure (Sagberg, 2002). The parameters that are used to evaluate safety measures within the engineering approach are primarily quantitative and relate to changes outside the actual driving experience. These parameters are often related to factors known to be associated with accident risk, such as driving speed, lane keeping, distance to traffic in front of car (Martens, Comte & Kaptein, 1997), collision frequency, and risk of injury or fatality (Elvik, Vaa & Østvik, 1989; Rothengatter, 2002).

One could argue that the engineering approach has been focussed on altering the physical characteristics of the road to increase traffic safety. The engineering approach is also concerned with the relation between the design of the road and how it affects driver behaviour. However, this interest lies mainly in how different physical safety measures affect choice of speed and lane keeping. In this respect, several projects have been started to examine how physical traffic safety measures affect such objective parameters (Martens et al., 1997; Sagberg, 2002). These studies have found a correlation between driving lane width and driving speed: wider roads lead to higher driving speed (Martens et al., 1997; Sagberg, 2002; Yagar & van Aerde, 1983).

The engineering approach to traffic safety has largely been a-theoretical. This means that engineering researchers often place less weight on a theoretical background to why there is a relation between speed and physical safety measures. When a theoretical background is given in an article or in a report (see Sagberg, 2002; Vaa & Bjørnskau, 2002), these theories are often risk models, such as Wilde's theory of risk homeostasis (Wilde, 1982; 1988) or Näätänen and Summala's (1976) zero-risk theory. These theories come from another division of traffic safety research which will be presented in the following.

2.2 Traffic psychology

The approach termed traffic psychology[1] by Rothengatter (2002) is primarily oriented towards a social science paradigm. This approach is a diverse field in itself, where several theoretical models are used to approach traffic safety from different perspectives. Among the most important models are the risk-perception theories of Wilde (1982; 1988) and Näätänen and Summala (1976) and also attitude-behaviour models such as Ajzen's theory of planned behaviour (Ajzen, 1985; Parker, Manstead & Stradling, 1995).

Among the phenomena studied in the traffic psychology approach are attitudes, expectations and risk-perception. These phenomena are often used to find motivational variables for driver behaviour, and to find predictors of aggressive or aberrant traffic behaviour. The predictors can then be used to find potential high-risk individuals (Parker, Lajunen & Stradling, 1998), who can then be offered behavioural treatment to reduce the risk of unwanted driver behaviour (Galovski, Blanchard, Malta & Freidenberg, 2003). Predictors for aberrant driving are usually linked with research on attitudes, using attitude-behaviour models such as the theory of planned behaviour as a theoretical basis (Ajzen, 1985; Parker et al., 1995; Rothengatter, 2002). The motivational variables (or 'extra motives', following Näätänen and Summala's (1976) terminology) can tell us which factors are risk factors in driving (Salminen & Lähdeniemi, 2002). This research is often founded on a theoretical background

---

[1] This approach circumvents the criticism presented by Groeger (2002), since Groeger's claim that a distinct school of traffic psychology does not exist only relates to the cognitive psychological approaches to traffic safety. As such, this paper is in agreement with Groeger (2002), that an independent cognitive traffic psychology does not exist, although other scientific traditions exist based on paradigms other than cognitive psychology that fit the description 'traffic psychology' (see Rothengatter, 2002).

of risk-perception or risk-estimation theories (Wilde, 1982; 1988; Näätänen & Summala, 1976).

Attitudes have often been studied using questionnaires (Parker et al., 1998; Yagil, 1998) or a combination of interviews, questionnaires and objective measures of driving speed (Haglund & Åberg, 2000). Risk perception and hazard perception have been studied through the use of questionnaires (Farrand & McKenna, 2001; Salminen & Lähdeniemi, 2002) and using response latency (Farrand & McKenna, 2001). The data generated from the questionnaires used in these studies are usually quantitative. The data generated from questionnaires with rating scales are a quantification of qualitative information. This means that a qualitative response to a question such as 'Do you agree/disagree with this statement?' is quantified by using a scale with numeric values.

An approach related to, but not specifically aimed at traffic safety, is the psychometric approach to risk perception (Slovic, 1987; for an overview, see also Boholm, 1998), personality subtypes of drivers (Ulleberg, 2002; Ulleberg & Rundmo, 2003) and the work related to finding factors underlying demands for risk mitigation (Sjöberg, 1999; 2000; Slovic, 1999). These approaches are more related to general risk research than traffic safety research. The issues which researchers in these fields work with are still important to traffic and transportation safety, as they deal with the determining factors of risk perception and estimation. The psychometric approach uses psychological scaling and multivariate analysis techniques, such as regression analysis (Sjöberg, 2000) and factor analysis (Boholm, 1998; Ulleberg, 2002; Ulleberg & Rundmo, 2003) to produce quantitative representations of risk perceptions and attitudes (Slovic, 1987). It has been found, for example that attitudes mediate a correlation between personality factors and driving behaviour (Ulleberg, 2002; Ulleberg & Rundmo, 2003) and that risk perception is mainly determined by the probability of harm, whereas it is the degree of seriousness of a consequence that affects demands for risk reduction (Sjöberg, 1999; 2000).

One can conclude from this description that 'traffic psychology' mainly uses quantitative methods and that the use of qualitative data is uncommon. Despite its focus on 'internal' phenomena such as attitudes, expectations and risk perception, this tradition has overlooked the experiences of the driver.

2.3 The psychophysical approach

The third and final major approach in traffic safety research is the psychophysical approach. This scientific domain attempts to find the physiological correlates of subjective phenomena such as fatigue (Lal & Craig, 2001), workload (Brookings, Wilson & Swain, 1996) and aggression (Galovski et al., 2003). The purpose of this approach is to better understand the psychology linked to such phenomena and through this find better ways of coping with the negative effects (Lal & Craig, 2001).

The methods used in this approach involve measuring physiological responses such as heart rate, systolic and diastolic blood pressure, galvanic skin response, eye blinking, saccadic movements, and respiration (Brookings et al., 1996; Galovski et al., 2003; Morris & Miller, 1996). Also, measurements may be taken of neural patterns in the central nervous system (CNS) using, for example, an electroencephalogram (EEG; Rokicki, 1995; Lal & Craig, 2001). In addition, use has been made of questionnaires (Milosevic, 1997), subjective reports (Lal & Craig, 2000), and rating scales such as NASA-TLX (Brookings et al., 1996) to measure the subjective or psychological aspects of fatigue. NASA-TLX is a rating scale that contains six subscales which measure mental, physical and temporal demand and performance, effort and frustration, and are scored from 0 (low) to 100 (high) (Hart & Staveland, 1988).

In common with both the engineering and traffic approaches, the methods and data used in the psychophysiological approach are mainly quantitative. The methods are often experimental in the sense that controlled circumstances and stimulation is used in conjunction with measures of psychophysiological activity (Brookings et al., 1996; Lal & Craig, 2001; Morris & Miller, 1996). This overview of the psychophysical tradition leads to the conclusion that quantitative data is used in abundance and that qualitative data is used to a lesser extent. Having presented a description of the general methods used and the types of data collected within traffic safety research, I will next give a summary of this paper's objectives.

2.4 Summary of the dominant perspectives within traffic safety research

In general, the use of quantitative data in traffic safety research is widespread and the preferred format. Quantitative data range from the measurement and calculation of objective data such as driving speed, lane keeping, and number of accidents (Martens et al., 1997; Sagberg, 2002), to questionnaire and survey studies of expectations, attitudes (Parker et al., 1998; Yagil, 1998) and risk perception (Slovic, 1987; Sjöberg, 2000), or the measurement of physiological correlates of fatigue (Lal & Craig, 2001), aggression (Galovski et al., 2003) and workload (Brookings et al., 1996).

A notable exception of the use of quantitative data was in a study conducted by Dorn and Brown (2003), where they interviewed police officers about their driver training, driving strategy and accident involvement. A narrative analysis of the interview transcripts was carried out and the researchers found that the police officers often regarded themselves as being highly aware of traffic hazards presented by other road users and that they had a tendency to minimize their own culpability and accident risk (Dorn & Brown, 2003).

From this overview it can be argued that there are rather few examples of experiential (or phenomenological) data in traffic safety research. This fact is contradictory to the claim made by many traffic safety researchers that the individual differences in experiences are important (e.g. experience of risk, Lal & Craig, 2001).

In this paper I will argue that a phenomenological understanding of driver experiences must precede the particular measures of given concepts. For example, in terms of the concept of risk compensation, Vaa and Bjørnskau (2002) state 'risk compensation is about how risk is experienced. If we are to understand what risk compensation is, we must understand the mechanisms that guide human experience of risk'. The challenge here is that actual human experience seems to be overlooked by the traditional quantitative measurements and constructs; phenomenology offers an alternative account of experience as being fluent and continuous, based in a temporal and embodied world that involves an organism that acts in and with an environment (Varelas, Thompson & Rosch, 1991). Beyond this philosophical stance, there is also a practical way to access experience. Simply put, in order to find out whether a person experiences (or perceives) risk or other individual phenomena, one must ask that person (Drury, 2002). The experience of risk can only be accessed from an internal perspective, such as by asking 'What is risky for me?'. From this it can be concluded that the data used to evaluate traffic safety measures are from 'outside' the driver's problems. Traditional, quantitative data arguably fail to grasp the most psychologically relevant aspects of the driving situation, namely the experiences of the driver.

This discrepancy between the goals and methods of traffic safety research suggests that there is a lack of knowledge concerning how people experience the driving situation and the different physical traffic safety measures. Hence, qualitative data may serve as a useful supplement to existing quantitative descriptions, whereby traffic research could be both quantitative (as in the aforementioned approaches) and qualitative. These two types of data could be viewed as different representations of the same phenomenon (Valsiner, 2000). In this respect, the different approaches should be understood as entailing different levels of knowledge regarding the phenomenon in question. What needs to be done is not to find a way to translate qualitative data into quantitative data (as done with some questionnaires) or vice versa, but rather to show how the qualitative and quantitative levels of description supplement each other.

3 Introducing a New Research Paradigm within Traffic Safety Research

The prime research subject in phenomenology as a philosophical tradition has been the world as it is experienced (Husserl, 1931; Merleau-Ponty, 1962). The phenomenological perspective on experienced worlds is in stark contrast to the philosophical heritages of Descartes and Plato that gave us rationalism and dualism. The dualistic presuppositions laid the basis for scientific positivism that is the leading paradigm within science today (Giorgi, 1970; Lynch, 1985). The phenomenological perspective changes these metaphysical presuppositions by going back to the phenomenon itself and it does this by describing objects as they are perceived and experienced. Phenomena are what we see, what we experience – they are to be described, and not constructed (Husserl, 1931).

Objectivity within phenomenology is achieved by researchers bracketing their theories, assumptions and ideologies. This makes it possible to perform analyses that do not question the existence of the research objects (Merleau-Ponty, 1962; Husserl, 1931). Phenomenologists stress the relation between the subject and the object (the human and its world), and rather than examining the two in isolation. Subject and object must be viewed as a unity, since human consciousness is 'intentional', meaning that consciousness (and behaviour) is directed toward something other than consciousness itself or behaviour itself. Neither can be understood, for instance, without taking into account the object towards which the activity was directed (Giorgi, 1983; Merleau-Ponty, 1962). This means that every experience has an object: to experience is to experience something (Varelas et al., 1991).

Phenomenologists attempt to understand the 'objective' in terms of subjectivity. This means that phenomenology studies what is given to consciousness as a person experiences it. This must not be regarded in a logical positivist way, but rather is to be viewed as a middle position where the 'objective' world of logical positivism is non-existent since the world to be perceived is always perceived by someone (Giorgi, 1983).

The human and its world are viewed as being in interaction, where neither the objects of the world nor the cognitive systems of the human solely determine the content of this interaction (Varelas, 1999). The Cartesian dualism between body and mind is rejected, since the human experiences the world through its body. The body is the subject of the personality; it is through the body that consciousness takes shape (Merleau-Ponty, 1962; Østerberg, 1994). It is the whole person that perceives and acts in the world. The person is not acting as a body made up of discrete parts (*partes extra partes*), but rather as a whole and integrated organism that functions as a unity (Merleau-Ponty, 1962).

3.1 Subjectivity is a valid research subject

The psychological (or experienced) world of an individual is situated between the subjective and the objective, and is not intersected by these concepts. For a human, the concepts of objective and subjective do not exist as detached. Each experience is as real as any other, as it exists for/with the person. Whether experiences are classified within the subjective or objective domain is not important. All objects and things are considered to be 'for-consciousness', because everything we know and feel must come through the consciousness (Giorgi, 1983). This means that consciousness is the psychological reality because it is 'the medium of access to whatever exists and is valid' (Gurwitsch, 1964, p. 166). The term 'psychological reality' is used to denote that the psychological world of an individual is different from the physical world. This is due to the intentionality of consciousness, where consciousness is directed towards objects and things outside itself, and it can be said that these objects transcend consciousness and that the consciousness emanates outwards into the world (Merleau-Ponty, 1962). In this perspective, subjectivity means that the human and its experience is a necessity for everything that is 'objective'. Subjectivity is just another part of the objective, since objectivity comes forth through human experience. This is based on the perspective of interrelations between the subjects and objects in the world. Whether other persons have direct access to this experience (as is demanded by logical positivism) does not disqualify them as psychological data in the phenomenological paradigm (Giorgi, 1970).

Subjectivity in the sense of 'what the individual experiences' can be researched using methods based on verbalizations of experiences, where the transcripts of the qualitative interview are condensed into meaning units (Giorgi, 1975; 1989; Kvale,

1996). A short description of the phenomenological method will be presented in the following.

3.2 The phenomenological method

Phenomenological analysis revolves around the search for essences and meanings. Analysis thus aims at condensing the meanings expressed by the subject into shorter, more structured formulations (Kvale, 1996). Before the researcher start to analyse the data, it is important that s/he brackets his/her foreknowledge and presuppositions. This means that the researcher should not question the realness of the phenomenon or allow the analysis to be guided by own assumptions, and should not question the existential nature of the phenomenon (Husserl, 1931; Merleau-Ponty, 1962). Rather, the researcher should strive to view the data as it is presented in the specific situation (Giorgi, 1998; Husserl, 1931). The method consists of five separate steps, each of which elaborates on the meaning of a description of a phenomenon (Giorgi, 1975; 1989; 1998; Kvale, 1996).

First, a description is needed. This description must be in the form of a text. It can be a transcript of an interview, a video-recorded conversation, or something similar. The *second* step is to read through the description to gain a sense of the whole. The *third* step is to locate the natural meaning units in the description. This involves dividing the description into smaller units that have a common meaning, or theme. Such a procedure is necessary, as it is impossible to carry out an analysis of something without breaking it up into parts. To do this, the researcher marks off where s/he finds a transition in the meaning of the text. Meaning units can be of different lengths, from parts of a paragraph to several paragraphs. When the meaning units is marked off, the researcher shifts from working with the interviewee's own words, to finding the core of each meaning unit. The *fourth* step implies a transformation of the subject's statements to find the implicit psychological knowledge contained in that description (Giorgi, 1989). The transition from meaning units to more condensed statements of psychologically relevant aspects of the phenomenon is called 'transformed meaning units'. The researcher tries to make explicit what is implicit, that is, what the essence of the experience is. The transformed meanings are often smaller in size than the original meaning units, and it can prove problematic to conserve the expressed meanings in the transformed units. The *fifth and final* step is to synthesize the transformed meanings to obtain the structure of the experienced phenomena. The result of this analysis is a short description, similar to a small history, and contains the essence of the phenomenon. This essence is what is supposed to be invariant in the experience of the phenomenon. All transformed meaning units should be present in the structured description.

When the analysis is finished, it is possible to look for common structures and variations in the structured descriptions of several subjects (Giorgi, 1998). This analysis can give an indication of what is common in the experiences of several subjects in relation to the given phenomenon. This analysis is of a higher level than the common quantitative methods since it describes the phenomena on a higher and more abstract level than the often-used numeric descriptions in, for example, traffic safety research.

3.3 The validation of qualitative data

What characterizes good phenomenological data? The phenomenological data cannot be evaluated using positivist criteria due to different metatheoretical perspectives, but should be evaluated using alternative qualitative research criteria. The general lack of a standardized research method makes this difficult to do in a procedural manner, although some criteria exist. These criteria are not supposed to be exhaustive but rather serve as guidance as to what is needed to have a valid approach to qualitative science.

The *first* criterion is freedom from bias (the bracketing employed by phenomenologists). This means that the researcher does not employing a given perspective when analysing the data. The *second* criterion is to establish intersubjective objectivity, which means that scientific data must be intersubjectively testable and reproducible. It can be difficult to perfectly reproduce the results from a qualitative interview, but with this kind of data the aim should rather be to see if two or more researchers can reach the same general understanding of the text (Kvale, 1996). *Third and finally*, the objectivity of the method can depend upon the relation to the phenomenon investigated, whether it expresses the real nature of the object or phenomenon studied. If the object studied is understood as linguistically and socially situated, the methods used in the social sciences can come closer to the research object than the methods used in the natural sciences. The relation between the data employed by the method and the nature of the object will affect the correctness of the method (Kvale, 1996).

3.4 The relation between quantitative and qualitative data

Qualitative data can be further validated by demonstrating relations between the essences from the phenomenological description of several subjects and changes in the quantitative data. If the analysis of several subjects reveals the same or similar phenomenological essences this can then be related to the 'objective' measures taken. Within traffic safety research, these measures can be driving speed, lane keeping, and psychophysiological measures of heart rate, blood pressure, and so forth. Changes in the psychophysical measures can then be checked to see whether they parallel the essences found in the phenomenological analysis. The same can be done with changes in driving speed or any of the quantitative measures taken in the given study.

Researchers have found a relation between subjective reports (qualitative) and psychophysiological (qualitative) measures (Lynch, 1985; Lal and Craig, 2000). Lal and Craig (2000) found that self-reports of fatigue paralleled psychophysiological correlates of fatigue, such as EEG (electroencephalograms), and heart rate. Other reports of the combination of qualitative and quantitative data come from studies of the psychotherapy of people with cardiovascular disease. Lynch (1985) showed that the function of the heart (systolic and diastolic blood pressure, heart rate) is affected by conversation. The procedures used in the experiments Lynch reported were continual recording of the blood pressure, heart rate and other cardiovascular measures during conversations or interviews with the patient concerning their treatment. A correlation between the changes in blood pressure and heart rate and the topic of conversation was found for several subjects; what was talked about clearly influenced the cardiovascular function of the patients (Lynch, 1985). This example concerned a therapeutic approach to the combination of qualitative (the conversation) and quantitative data (blood pressure), but it suggest how these two forms of data can be used to improve the understanding of medical or psychological research problems.

4 The application of the phenomenological interview to evaluate drivers' experiences

It is generally difficult to obtain verbal data from a person who engages in activities that involve a high cognitive workload. Examples of this type of situation are multi-tasking situations where two or more tasks tap the same mental resources (Baddeley, 1987), or where a person has to navigate an unfamiliar terrain while also trying to avoid collision with moving objects. Common instances of these types of multi-tasking situations are conversions with hand-held mobile phones during driving, something that has been banned in several countries due to the extra demand on attention resources that can interfere with driving (Patten, Kircher, Östlund & Nilsson, 2004; Recarte & Nunes, 2003). It is also problematic that the situations where the operator (the driver) has to work the most to maintain control – which are the most interesting ones from a human-factors perspective – also are the situations where it is most difficult to obtain verbal reports from the drivers about what information they are seeking and their reasoning and information gathering.

However, these shortcomings can be somewhat compensated for if phenomenological interviews are used in combination with controlled experiments, such as those performed in simulated environments (Bjørkli, Hoff, Jenssen, Øritsland & Ulleberg, this volume; see Farmer, van Rooij, Riemersma, Jorna & Moraal, 2003 and Hettinger & Haas, 2003; for an overview of elements in simulator-based training). Simulated environments fit very well with phenomenological interviews in that they ensure that each participant observes and acts in the same scenarios. Hence, we can be quite sure that the tasks the research participants encounter are of similar nature, something that would simplify the comparison of the phenomenological essences as one experimentally controls for the effect of context and task characteristics in relation to the driver's experience of task and safety measures. The use of staged or simulated environments additionally allow for the collection of other 'objective' data such as vehicle dynamics and psychophysical responses of participants in experiments. These data can then be compared with the transcripts and derived phenomenological essences in order to create an understanding of the research object from both qualitative and quantitative representations of the data.

As mentioned, it is common knowledge that drivers tend to talk less when they encounter demanding situations. This is may be because speaking uses mental resources (Reisberg, 2001, Recarte & Nunes, 2003; Baddeley, 1987), resources which in demanding situations must be used to tackle the task by focusing attention and activity towards the problematic aspects of the situation. This shortcoming of direct online verbal reports during task performance can be (partly) compensated by the use of video cameras that make online recordings from different viewpoints of the situation (for methodological and theoretical aspects of the use of video cameras in social science research see Laws & Barber, 1989; Hall, 2000) and that allow for a perspective that capture the same elements that the experimental participants see. In normal everyday situations the choice of camera angle and overview is very difficult and also highly problematic in that we do not know where the participants will look for information. However, staged or simulated environments allow for a good measure of control over the events that occur. These also allow the researcher some way of predicting what type of information the participant in the experiment will seek at given points in time, and hence also allow them to anticipate the participant's behaviour (Neisser, 1987). This knowledge may be used to guide the location of the cameras. For the ease of reading, it an overview of the following aspects should be obtained: the general task environment (e.g. simulated or staged traffic environment); the driver's head and body orientation and the use of important interface controllers. However, it is not the purpose of this paper to discuss the theory and use of video recordings in the social sciences, and in-depth reviews of the use of video for research purposes can be found elsewhere (see e.g. Laws & Barber, 1989; Hall, 2000).

4.1 The use of video recordings to guide and contextualize the phenomenological interview

Video recordings form the basis of phenomenological interviews. The research participant and the interviewer view the video recording(s) and the subject will then be asked to describe in more detail what s/he experiences in given situations. By using videotapes as the basis for interviews, subjects (drivers) have as much time as they wish to describe their experiences of driving.

During the interview the subject will be asked to identify incidents on the video that s/he experienced as significant or important. Such events will then be used to later divide the interview transcript into several 'separate' phenomenological analyses. The researcher can also ask the interviewee to describe events that the researcher finds interesting, and these events can also be used to divide the transcript into several parts, with each part relating to a given episode. To facilitate such an event-based phenomenological analysis, the researcher should consider also video recording the interview session in a way that allows a view of the interviewer, the research participant and the pre-recorded video of task performance in the simulated environment.

Other possibilities could be to use the psychophysiological data to divide the driving situation into 'episodes'. Since the transcript and the following three steps of the phenomenological analysis (Giorgi, 1989; 1998; Kvale, 1996) are readily available, it is possible to create condensed meaning essences that correspond to the time-dependent change in psychophysical data.

4.2   Use of the video recording during the analysis of interview data

After the researcher has created the meaning units, the video recordings of driving sessions and the interview can be incorporated into the analysis. This is an exception from the 'standard' phenomenological analysis (Giorgi, 1989; 1998), but it is necessary to do this in order to be able to map the relationship between the changes in the driving environment and the driver's experience of these changes. The video recording of the phenomenological interview shows the participant that watches the video of herself/himself driving while talking about her/his experience and reasoning that occurs during the situations seen on the video. This allow for a way to combine and integrate the transcription data with the changes in the traffic scene.

Before the fourth and fifth steps of the phenomenological analysis the researcher should determine which parts of the transcripts fit with which parts in the video recording of the driving situation. By doing this the researcher can isolate the meaning units 'belonging' to the change in the simulator environment, and steps four and five can be done as though several transcripts exist, one for each event in the driving simulator that the subject reported as important (or the event that the interviewer found interesting). This use of the phenomenological analysis can be termed an episode-based phenomenological analysis, where a whole transcript of a conversation concerning a process, such as driving, is divided into episodes. These episodes can then be analysed separately, or all taken together (as in a standard analysis).

4.3   Checking for parallels between the qualitative and the quantitative data

When employing several subjects in this research design, it is possible to compare the essences of these subjects for the same events. Similarities or differences in these essences can then be presented. It is also possible to carry out a standard phenomenological or a content analysis (Neuendorf, 2002) of the transcript (in addition to the episode-based analysis) and then to compare the results of one analysis of the whole driving situation with the other analyses.

The most important aspect of this design is the possibility to compare qualitative and quantitative data. This can be done to see whether there are consistent changes in driving speed or in psychophysiological data that are related to particular experiences as the persons describe them. This can be done by asking questions such as: 'Is there a correlation between the heart rate and the perception of task difficulty, and are this correlation also paralleled by a particular type of experiential description?' 'If the qualitative descriptions from several subjects are similar in some episodes, is there also a similar trend in the change in the quantitative data for these individuals?'

It would also be interesting to check how drivers perceive and experience a physical traffic safety measure that is presented in the simulation scenario. This could give a new approach to the evaluation of traffic safety measures. If several subjects experience an increase in danger or risk and reduce their speed accordingly (this might be measured by finding similarity between essences during the episode and the consistent reduction of speed by the same subjects), it could give new information with regard to risk theories by identifying under which circumstances risk is experienced. If a safety measure leads to an increased experience of danger together with increases in, for example, heart rate and blood pressure, it can be assumed that the safety measure works by inducing anxiety-like symptoms. In this case, it is not the examples that can validate the method, but rather the data that emerge when this method is employed. The statements made here are just some possible ways of comparing the two types of data; an exhaustive presentation can only be made with the help of the imagination of the researcher.

4.4   Pros and cons of the research methodology

One possible advantage using video recording as a basis for the qualitative interview is that the driver will have unlimited time to present his understanding and experiences of the driving situation. It is possible to obtain 'thick' descriptions of a given situation that are applicable to an actual 'objective' change in the scenery. The experimenter and the subject can then stop the film or rewind it if they feel that a given situation is of interest. Another advance with this method is that the commentary can serve as a cue for recollection of the given experiences. For instance, the subject may say, 'When I said that, I felt like *that*, I experienced ...', and so forth.

The drawbacks with holding the interview afterwards (and not during driving) are that the person has to recollect the experiences from memory. Since much of driver behaviour is automated, it is possible that the subject will not have any conscious recollection of the given moments presented on the film. It is also a possibility that the subjects may confuse different experiences, especially if the scenario takes a long time to drive through. However, it is assumed that the possibility for drivers to talk more when using the video-based interview (with no limit on time) outweighs the more direct approach of holding the interview in the car while driving. The use of video recordings also appears to makes it easier to divide interviews into episodes, compared to when interviews occurs while driving.

The general use of both qualitative and quantitative data in this design makes it a versatile research approach, since it gathers several types of data that each complements the understanding of the driving situation and the driver's experience. The qualitative part presented in this paper is also an unobtrusive way of performing qualitative science, as it does not reduce the validity of the quantitative data. Furthermore, the use of video cameras reduces the contact between the researcher and the research participant. This way of preparing for a qualitative interview probably does not reduce the research subject's focus on the tasks presented to him/her, as the social influence of the researchers can be kept to a minimum during the experimental stage.

4.5   Limitations of the suggested methodology

Some limitations of the presented design must also be considered. The combination of the experimental setting and the phenomenological analysis is best fitted to describing traffic safety measures that appear to the driver while s/he is driving. The presented method will then fit best to evaluations of the traffic safety measures made by the engineering approach, as the measures implemented by this tradition are tangible and often include visible changes in the design of the road. However, the methods may also be used to investigate what affects the driver's perception of speed, for example by changing road infrastructure through the removal of objects, trees and buildings close to the road, the introduction of visual illusions (Sagberg, 2002), or changes in the 'optic flow field' to alter the experience of movement (Gibson, 1979; Owen & Warren, 1987). The phenomenological approach may further be applied to study the phenomenon of speed adaptation, building on the notion that the experience of speed is reduced after prolonged exposure to high speed (Mashour, 1981).

5 Discussion and Conclusions

Traffic safety research been mostly occupied with quantifying aspects of the driving situation and the possible consequences of types of driver behaviour. This preoccupation with qualitative data is probably due to the hegemony of qualitative methods and adhering philosophical stance. The possibility of contrasting and supplementing the traditional approaches with a qualitative approach to traffic safety research has been discussed by outlining a research strategy based upon a combination of the experimental and the phenomenological method (Giorgi, 1989; 1998). This method is of a generic type and its principles are not limited to a particular subset of tasks or research objects (such as traffic safety measures). The actual suitability of this method will depend upon the employment of the given design. It is more difficult to show how qualitative methods can improve traffic safety, as qualitative research in this field to date is scarce. Although one article using only qualitative data has presented ways of altering police driver training in Great Britain (Dorn & Brown, 2003), it shows that qualitative methods can be an interesting approach to certain traffic safety problems.

The possible improvement in traffic safety with the use of new epistemological approaches and methods is hard to foresee for two reasons: first, the results of future research are impossible to know in advance; second, the currently used paradigms and epistemological assumptions can conceal the possible advantages by employing new epistemological tools. Having said this, it should also be pointed out that the big steps forward in science have been made when researchers and theorists have gone beyond the ruling paradigms and established new theorems, insights and methods (Valsiner, 2000). The innovative scientific discoveries made, for example, by Copernicus, Einstein and Bohr were made possible only by breaking with the dominating scientific paradigm of the time (Feyerabend, 1975; Schaanning, 2000). This lesson from history shows us that is can be interesting to try out new ways of accumulating knowledge, even though new approaches often meet considerable resistance in the respective research communities (Kuhn, 2002). Some researchers have also claimed that the occupation and use of positivist-oriented methods within areas such as work psychology have led to rigidity in the use of methods and what data are recorded (Johnson & Cassell, 2001). To prevent such rigidity in traffic safety research (if one has not already occurred), researchers should look to new ways of approaching traffic safety. This can be done by being aware of and critical towards the epistemological and philosophical basis for the employed methods and the nature of the examined object.

Using an episode-based phenomenological analysis that maps changes in the environment with verbalized experiential data can give us clues as to why traffic measures actually work the way they do. These data are related to the experience of the driver and through that they will yield information about intrinsic (internal) aspects of the driving situation. The suggested design presents a new level of data that to date has not been used within traffic safety research. The experiential aspects of driving and other related activities could be researched using a method like the one presented here. The combination of existing methods and other qualitative approaches might improve our understanding of driver behaviour. New aspects of the driving situation become available and parallels between the 'inner' and 'outer' world of driving and traffic behaviour can be found and through this affect road safety on a level that is present almost all the time – namely the human body/mind. The acceptance and use of new research paradigms and their related methods might be a new way forward for traffic safety research.

The challenge is to explore and/or transcend the boundaries of the ruling paradigms and in some areas move into uncharted territory (see Giorgi, 1970, for a lengthy discussion of this topic within psychology). The use of new methods does not need to compete with established perspectives and paradigms. Instead, they can be used to complement and fill out the description and knowledge relating to traffic safety and other research areas.

References

Ajzen, I. (1985). From intentions to actions: A theory of planned behavior. In J. Kuhl, & J. Beckman (Eds.), *Action-control: From cognition to behavior*. Heidelberg: Springer Verlag.

Baddeley, A. (1987). *Working Memory*. Oxford, UK: Oxford University Press.

Boholm, Å. (1998). Comparative studies of risk perception: a review of twenty years of research. *Journal of Risk Research*, 1, 135-163.

Brookings, J. B., Wilson, G. F. & Swain, C. R. (1996). Psychophysiological responses to changes in workload during simulated air traffic control. *Biological Psychology*, 42, 361-377.

Dorn, L. & Brown, B. (2003). Making sense of invulnerability at work – a qualitative study of police drivers. *Safety Science*, 41, 837-859.

Drury, C. G. (2002). Measurement and the practicing ergonomist. *Ergonomics*, 45, 988-990.

Elvik, R., Vaa, T. & Østvik, E. (1989). *Trafikksikkerhetshåndboka* [Handbook of Traffic Safety]. Oslo; Norway: Transportøkonomisk institutt.

Farmer, E., van Rooij, J., Riemersma, J., Jorna, P., & Moraal, J. (2003). *Handbook of Simulator-Based Training*. Aldershot, England: Ashgate Publishing.

Farrand, P. & McKenna, F. P. (2001). Risk perception in novice drivers: the relationship between questionnaire measures and response latency. *Transportation Research Part F*, 4, 201-212.

Feyerabend, P. (1975). *Against Method*. London, UK: Verso.

Galovski, T. E., Blanchard, E. B., Malta, L. S. & Freidenberg, B. M. (2003). The psychophysiology of aggressive drivers: comparison to non-aggressive drivers and pre- to post-treatment change following a cognitive-behavioural treatment. *Behaviour Research and Therapy*, 41, 1055-1067.

Gibson, J. J. (1979). *The Ecological Approach to Visual Perception*. London, UK; Hougthon Mifflin Company.

Giorgi, A. (1970). *Psychology as a Human Science: A Phenomenogically Based Approach.* New York; Harper & Row, Publishers.

Giorgi, A. (1975). An application of the phenomenological method in psychology. In A. Giorgi, C. Fischer, & E. Murray, (Eds.), *Duquesne studies in phenomenological psychology,* II. (pp. 82-103). Pittsburgh, PA: Duquesne University Press.

Giorgi, A. (1983). Concerning the possibility of phenomenological psychological research. *Journal of Phenomenological Psychology,* 14, 129-169.

Giorgi, A. (1989). One type of analysis of descriptive data: Procedures involved in following a scientific phenomenological method. *Methods,* 39-61.

Giorgi, A. (1998). *Critical thinking under a phenomenological perspective.* Paper presented at the Psychological institute, Norwegian University of Science and Technology, Trondheim, 21. sept. 98, to 25 sept. 98.

Groeger, J. A., & Rothengatter, J. A. (1998). Traffic psychology and behaviour. *Transportation Research Part F,* 1, 1-9.

Gurwitsch, A. (1964). *The Field of Consciousness.* Pittsburgh: Duquesne University Press.

Haglund, M, & Åberg, L. (2000). Speed choice in relation to speed limits and influences from other drivers. *Transportation Research Part F,* 3, 39-51.

Hall, R. (2000). Video recording as theory. In D. Lesh & A. Kelley (Eds.), *Handbook of Research Design in Mathematics and Science Education.* (pp. 647-664). Mahwah, NJ: Lawrence Erlbaum.

Hart, S. G. & Staveland, L. E. (1988). Development of the NASA-TLX (Task Load Index): results of empirical and theoretical research. In P. Hancock & N. Meshkati (Eds.) *Human Mental Workload.* (pp. 139-183). Amsterdam; Elsevier.

Hettinger, L. J., & Haas, M. W. (2003). *Virtual and Adaptive Enviornments.* Mahwah, NJ: Lawrence Erlbaum.

Hills, B. L. (1980). Vision, visibility, and perception in driving. *Perception,* 9, 183-216.

Husserl, E. (1931). *Ideas: General Introduction to Pure Phenomenology.* (transl. W. R. Boyce Gibson). London, UK: Allen & Unwin, Ltd.

Johnson, P. & Cassell, C. (2001). Epistemology and work psychology: New agendas. *Journal of Occupational and Organizational Psychology,* 74, 125-143.

Kuhn, T. S. (2002). *Vitenskapelige revolusjoners struktur* [The Stucture of Scientific Revolutions, 2. edition] (transl. Lars Holm-Hansen). Oslo; Norway: Spartacus Forlag AS.

Kvale, S. (1996). *InterViews: An Introduction to Qualitative Research Interviewing.* Thousand Oaks, CA: Sage Publications, Inc.

Lakoff, G. & Johnson, M. (1999). *Philosophy in the Flesh: The Embodied Mind and its Challenge to Western Thought.* New York, NY; Basic Books.

Lal, S. K. L. & Craig, A. (2000). Driver fatigue: Psychophysiological effect. *The Fourth International Conference on Fatigue and Transportation,* Australia.

Lal, S. K. L. & Craig, A. (2001). A Critical Review of the Psychophysiology of Driver Fatigue. *Biological Psychology,* 55, 173-194.

Laws, J. V., & Barber, P. J. (1989). Video analysis in cognitive ergonomics: a methodological perspective. *Ergonomics,* 32(11), 1303-1318.

Lynch, J. J. (1985). *The Language of the Heart: The Human Body in Dialogue.* New York, Basic Books, Inc. Publishers.

Martens, M., Comte, S., & Kaptein, N. (1997). *The Effects of Road Design on Speed Behaviour: A Literature Review.* (Contract no. RO-96-SC.202), Espoo, Finland: VTT Communities & Infrastructure.

Mashour, M. (1981). The Information Basis for the Perception of Velocity. *Acta Psychologia,* 48, 69-78.

McKenna, F. P. (1993). It won't happen to me: Unrealistic optimism and the illusion of control. *British Journal of Psychology,* 84, 39-50.

Merleau-Ponty, M. (1962). *Phenomenology of Perception.* London; England: Routledge & Kegan Paul.

Morris, T. L. & Miller, J. C. (1996). Electrooculographic and performance indices of fatigue during simulated flight. *Biological Psychology,* 42, 343-360.

Näätänen, R & Summala, H. (1976). *Road user behaviour and traffic accidents.* New York; Elsevier.

Neisser, U. (1987). From direct perception to conceptual structure. In U. Neisser (ed.), *Concepts and Conceptual Development: Ecological and Intellectual Factors in Categorization* (pp. 11- 24) Cambridge, UK; Cambridge University Press.

Neuendorf, K. A. (2002). *The Content Analysis Guidebook.* Thousand Oaks, CA: Sage Publications.

Owen, D. H., & Warren, R. (1987). Perception and control of self-motion: Implications for visual stimulation of vehicular locomotion. I L. S. Mark, J. S. Warm, & R. L. Huston (Red.) *Ergonomics and Human Factors: Recent Research*, (s. 40-70). New York: Springer-Verlag.

Parker, D., Manstead, A. S. R. & Stradling, S. (1995). Extending the theory of planned behavior: The role of personal norm. *British Journal of Social Psychology*, 34, 127-137.

Parker, D., Lajunen, T. & Stradling, S. (1998). Attitudinal predictors of interpersonally aggressive violations on the road. *Transportation Research Part F*, 1, 11-24.

Patten, C. J. D., Kircher, A., Östlund, J., & Nilsson, L. (2004). Using mobile telephones: cognitive workload and attention resource allocation. *Accident Analysis & Prevention*, 36(3), 341-350.

Recarte, M. A., & Nunes, L. M. (2003). Mental workload while driving: Effects of visual search, discrimination and decision making. *Journal of Experimental Psychology: Applied*, 9(2), 119-137.

Reisberg, D. (2001). *Cognition: Exploring the Science of the Mind.* London, UK: W. W. Norton & Company.

Rokicki, S. M. (1995). Psychophysiological measures applied to operational test and evaluation. *Biological Psychology*, 40, 223-228.

Rothengatter, T. (2002). Driver's illusions – no more risk. *Transportation Research Part F*, 5, 249-258.

Sagberg, F. (2002). *Påvirkning av bilførere gjennom utforminingen av veisystemet. Del II: Vegutforming og kjørehastighet.* [Influencing drivers through design of the road system. Part II, Road design and driving speed]. Draft paper TØI Rapport. Oslo; Transportøkonomisk Institutt.

Salminen, S. & Lähdeniemi, E. (2002). Risk factors in work-related traffic. *Transportation Research Part F*, 5, 77-86.

Schaanning, E. (2000). *Modernitetens oppløsning* [The dissolution of moderninsm]. Oslo, Norway: Spartacus Forlag AS.

Sjøberg, L. (1999). Consequences of perceived risk: Demand for mitigation. *Journal of Risk Research*, 2, 129-149.

Sjøberg, L. (2000). Consequences matter, 'risk' is marginal. *Journal of Risk Research*, 3, 287-295.

Slovic, P. (1987). Perception of Risk. *Science*, 236, 280-285.

Slovic, P. (1999). Comment: Are trivial risks the greatest risk of all? *Journal of Risk Research*, 2, 281-288.

Statistisk Sentralbyrå (2006). *Personer drept eller skadd, etter trafikantgruppe og skadegrad.* Can be found at http://www.ssb.no/emner/10/12/20/vtuaar/tab-2007-06-11-02.html [2008-02-26].

Ulleberg, P. (2002). Personality subtypes of young drivers. Relationship to risk-taking preferences, accident involvement, and response to a traffic safety campaign. *Transportation Research Part F*, 4, 279-297.

Ulleberg, P. & Rundmo, T. (2003). Personality, attitudes and risk perception as predictors of risky driving behaviour among young drivers. *Safety Science*, 41, 427-443.

Vaa, T., & Bjørnskau, T. (2002). *Fart, følelser og risiko: Drøfting av indre mekanismer ved bilføreres fartsvalg.* [Speed, emotions and risk: Discussion of inner mechanisms at driver's speed choice]. TØI Rapport 607/2002. Oslo; Transportøkonomisk Institutt

Valsiner, J. (2000). Data as representations: Contextualizing qualitative and quantitative research strategies. *Social Science Information*, 39, 99-113.

van Lennep, D. J. (1987). The psychology of driving a car. In J. J. Kockelmans (Ed.), *Phenomenal Psychology*, (pp. 217-227), Dordrecht, The Netherlands: Martinus Hijhoff Publishers.

Varelas, F. J. (1999). Steps to a Science of Inter-Being: Unfolding the Dharma Implicit in Modern Cognitive Science. In G. Watson, S. Batchelor & G. Claxton (Eds.). *The Psychology of Awakening*, (pp. 71-89). London, UK: Rider.

Varelas, F. J., Thompson, E., & Rosch, E. (1991). *The Embodied Mind: Cognitive Science and Human Experience.* Cambridge; Massachusetts: The MIT Press.

Wilde, G. J. S. (1982). The theory of risk homeostasis: Implications for safety and health. *Risk Analysis*, 2, 209-258.

Wilde, G. J. S. (1988). Risk homeostasis and traffic accidents: propositions, deductions and discussion of recent commentaries. *Ergonomics*, 31, 441-468.

Wittgenstein, L. (1953/1997). *Filosofiske Undersøkelser.* (Transl. by Mikkel B. Tin). Oslo: Pax Forlag A/S.

Yagil, D. (1998). Gender and age-related differences in attitudes toward traffic laws and traffic violations. *Transportation Research Part F*, 1, 123-135.

Yagar, S., & van Aerde, M. (1983). Geometric and environmental effects on speeds of 2-lane highways. *Transportation Research*, 17A, 315-325.

Østerberg, D. (1994). Innledning. [Introduction]. In M. Merleau-Ponty, *Kroppens fenomenologi* (pp. V-XII). Oslo, Norway: Pax Forlag A/S.